United States
Department of
Agriculture

Forest Service

Pacific Northwest
Research Station

General Technical Report
PNW-GTR-779
March 2009

# Ecological Foundations for Fire Management in North American Forest and Shrubland Ecosystems

J.E. Keeley, G.H. Aplet, N.L. Christensen, S.G. Conard, E.A. Johnson, P.N. Omi, D.L. Peterson, and T.W. Swetnam

The Forest Service of the U.S. Department of Agriculture is dedicated to the principle of multiple use management of the Nation's forest resources for sustained yields of wood, water, forage, wildlife, and recreation. Through forestry research, cooperation with the States and private forest owners, and management of the National Forests and National Grasslands, it strives—as directed by Congress—to provide increasingly greater service to a growing Nation.

The U.S. Department of Agriculture (USDA) prohibits discrimination in all its programs and activities on the basis of race, color, national origin, age, disability, and where applicable, sex, marital status, familial status, parental status, religion, sexual orientation, genetic information, political beliefs, reprisal, or because all or part of an individual's income is derived from any public assistance program. (Not all prohibited bases apply to all programs.) Persons with disabilities who require alternative means for communication of program information (Braille, large print, audiotape, etc.) should contact USDA's TARGET Center at (202) 720-2600 (voice and TDD). To file a complaint of discrimination, write USDA, Director, Office of Civil Rights, 1400 Independence Avenue, SW, Washington, DC 20250-9410 or call (800) 795-3272 (voice) or (202) 720-6382 (TDD). USDA is an equal opportunity provider and employer.

## Authors

**J.E. Keeley** is a research ecologist, U.S. Department of the Interior, Geological Survey, Western Ecological Research Center, Sequoia-Kings Canyon Field Station, Three Rivers, CA 93271, and adjunct professor, Department of Ecology and Evolutionary Biology, University of California, Los Angeles, CA 90095; **G.H. Aplet** is a senior forest ecologist, The Wilderness Society, 1660 Wynkoop Street, Suite 850, Denver, CO 80202; **N.L. Christensen** is a professor, Duke University, Nicholas School of the Environment and Earth Sciences, Durham, NC 27708; **S.G. Conard** is a national program leader, U.S. Department of Agriculture, Forest Service, Vegetation Management and Protection Research, Rosslyn Plaza C, 4th Floor, 1601 North Kent Street, Arlington, VA 22209; **E.A. Johnson** is a professor, University of Calgary, Department of Biological Sciences, Calgary, AB, T2N 1N4 Canada; **P.N. Omi** is a professor emeritus, Colorado State University, Warner College of Forest Resources, Western Forest Fire Research Center, Fort Collins, CO 80523-1472; **D.L. Peterson** is a biological scientist, U.S. Department of Agriculture, Forest Service, Pacific Northwest Research Station, Pacific Wildland Fire Sciences Laboratory, 400 N 34th Street, Suite 201, Seattle, WA 98103; and **T.W. Swetnam** is a professor, University of Arizona, Laboratory of Tree-Ring Research, Tucson, AZ 85731.

Cover: Prescribed fire in Giant Forest, Sequoia National Park. Photo by Eric Knapp.

# Abstract

**Keeley, J.E.; Aplet, G.H.; Christensen, N.L.; Conard, S.C.; Johnson, E.A.; Omi, P.N.; Peterson, D.L.; Swetnam, T.W. 2009.** Ecological foundations for fire management in North American forest and shrubland ecosystems. Gen. Tech. Rep. PNW-GTR-779. Portland, OR: U.S. Department of Agriculture, Forest Service, Pacific Northwest Research Station. 92 p.

This synthesis provides an ecological foundation for management of the diverse ecosystems and fire regimes of North America, based on scientific principles of fire interactions with vegetation, fuels, and biophysical processes. Although a large amount of scientific data on fire exists, most of those data have been collected at small spatial and temporal scales. Thus, it is challenging to develop consistent science-based plans for large spatial and temporal scales where most fire management and planning occur. Understanding the regional geographic context of fire regimes is critical for developing appropriate and sustainable management strategies and policy. The degree to which human intervention has modified fire frequency, intensity, and severity varies greatly among different ecosystems, and must be considered when planning to alter fuel loads or implement restorative treatments. Detailed discussion of six ecosystems—ponderosa pine forest (western North America), chaparral (California), boreal forest (Alaska and Canada), Great Basin sagebrush (intermountain West), pine and pine-hardwood forests (Southern Appalachian Mountains), and longleaf pine (Southeastern United States)—illustrates the complexity of fire regimes and that fire management requires a clear regional focus that recognizes where conflicts might exist between fire hazard reduction and resource needs. In some systems, such as ponderosa pine, treatments are usually compatible with both fuel reduction and resource needs, whereas in others, such as chaparral, the potential exists for conflicts that need to be closely evaluated. Managing fire regimes in a changing climate and social environment requires a strong scientific basis for developing fire management and policy.

Keywords: Fire ecology, fire hazard, fire regime, fire risk, fire management, fuels, fuel manipulation, prescription burning, restoration.

## Summary

This review uses a scientific synthesis to provide an ecological foundation for management of the diverse ecosystems and fire regimes of North America. This foundation is based on the following principles that inform management of fire-affected ecosystems:

- Potential future management options and goals need to be consistent with current and past fire regimes of specific ecosystems and landscapes and be able to anticipate and adjust to future conditions.
- The effects of past management activities differ among ecosystems and fire regime types.
- Differences in fire history and land use history affect fuel structures and landscape patterns and can influence management options, even within a fire regime type.
- The relative importance of fuels, climate, and weather differs among regions and ecosystems within a region; these differences greatly affect management options.
- Plant species may be unable to adapt to alterations in fire regimes.
- The effects of patch size must be evaluated within the context of fire regime and ecosystem characteristics.
- Fire severity and ecosystem effects are not necessarily correlated.
- Appropriate options for fuel manipulations differ within the context of vegetation structure, management objectives, and economic and societal values.
- Fuel manipulations alter fire behavior but are not always reliable barriers to fire spread.
- Understanding historical fire patterns provides a foundation for fire management, but other factors are also important for determining desired conditions and treatments.

Several challenges exist for implementing these principles in contemporary fire management. Although a large amount of scientific data on fire exists, most of those data have been collected at fine spatial and short temporal scales, whereas most of the potential issues and applications of those data are at broad and long-term scales. Basing decisions and actions on these data often requires extrapolation to different scales and different conditions, such that error can be introduced in the process. In addition, most land management organizations operate according to many legal and regulatory mandates, some of which are compatible with ecologically based fire management and some of which constrain potential options. Finally, a warming climate and other dynamic changes in the biological, physical, and social environment are introducing new sources of complexity and uncertainty that influence strategic planning and day-to-day activities.

Sustainable ecosystem-based management, which is now the standard on most public lands, will be successful only if fire policy and management are (1) based on ecological principles, (2) integrated with other resource disciplines (wildlife, hydrology, silviculture, and others), and (3) relevant for applications at large spatial and temporal scales. Fire is such a pervasive disturbance in nearly all ecosystems that failure to include it as part of managing large landscapes will inevitably lead to unintended outcomes.

# Contents

- 1 **Introduction**
- 3 Ponderosa Pine (Western United States)
- 3 Chaparral (Pacific South Coast)
- 5 Boreal Forest (Alaska and Canada)
- 6 Great Basin Sagebrush (Intermountain West)
- 7 Pine and Pine/Hardwood Forests (Southern Appalachians)
- 8 Longleaf Pine (Southeastern United States)
- 9 **Fire Regimes as a Framework for Understanding Processes**
- 9 Fuel Consumption and Fire Spread Patterns
- 12 Fire Intensity and Severity
- 14 Fire Frequency
- 17 Fire Patch Size and Distribution
- 19 Fire Seasonality
- 19 **Climate and Weather Effects on Fire Regime**
- 20 Climate and Fire Activity
- 22 Fire Weather
- 24 **Biogeographical Patterns of Fire Regimes**
- 27 Recent Changes in Fire Regimes
- 32 Human Impacts on Fire Regimes
- 35 Effects of Fire Exclusion on Forest and Shrubland Structure
- 43 Effectiveness of Fire Suppression
- 45 Fire Management and Ecosystem Restoration
- 46 Effectiveness of Prescription Burning
- 50 Effectiveness of Mechanical Fuel Manipulations
- 53 Ecosystem Effects of Mechanical Harvesting Versus Fire
- 53 Applications in Science-Based Resource Management
- 57 **Acknowledgments**
- 57 **English Equivalents**
- 57 **Common and Scientific Names**
- 58 **Literature Cited**

# Introduction

**This paper places the role of fire in a framework that will inform fire management of ecosystems at different spatial and temporal scales.** Although we focus on North America, the concepts discussed here have broader application. Fire occurs in most North American ecosystems, and most of these systems are resilient to fires that occur within a broad range of variability in frequency and intensity. Fire has been influenced by humans since before European settlement. On some landscapes, human impacts have resulted in widespread disruption of historical fire regimes and placed ecosystems on a trajectory leading to a less stable and less sustainable future. This scenario can have profound impacts on human social and economic systems as well as on the natural resources that provide us with numerous tangible and intangible benefits. As human presence has increased, there has been a concomitant increase in property and other values that are potentially at risk from unintended fire and in the perceived need to manage fire to reduce those risks. Many "natural" ecosystems (box 1) are also threatened by past and present fire management and land management practices.

**We show the diverse roles fire has played in different ecosystems, necessitating a regional approach to fire management, at least partially in response to human effects through fire exclusion in some cases and increased fire occurrence in other cases; ecosystem-based management requires different strategies on different landscapes.** We also focus on the relative role of different land management practices on fuel accumulation and fire hazard. Fire suppression is only one factor leading to increased fire hazard, and has not changed fire hazard in all ecosystems. Furthermore, land management activities such as logging and grazing, which some assume have reduced fire hazard, have actually exacerbated fire hazards on some landscapes. We also discuss how climatic variability and change are expected to alter future fire regimes and the potential impact of management responses to these changes. Finally, we examine regions where expanding urban populations have resulted in large portions of human settlements being exposed to high fire danger and altered local management options.

We begin by briefly describing examples of fire and fire management effects in six ecosystems. These examples illustrate the complexity of many fire issues, and the need for fire management that reflects the complexities of North American ecosystems and their different relationships to fire.

---

**Fire suppression is only one factor leading to increased fire hazard, and has not changed fire hazard in all ecosystems.**

*Box 1.*
## What Is Natural?

Referring to a place or process as "natural," ecologists most often mean "absent of human influence," which is the meaning intended for "natural" in this paper, although "limited" human impact may be a more realistic goal. This is not to dismiss or even participate in the dialogue about the relationship of humans and nature. Rather, we do this out of need for a word to describe a baseline frame of reference for understanding human influences. The tradition of using "natural" in this manner is well established, and no other word seems to fit the intent.

Over the past few millennia, only the more remote places in North America could be said to have been in this sense natural, and this is particularly true with respect to the occurrence and behavior of fire. Humans have used fire for most of their existence to modify and manage their environments (Pyne 1982, 2001), and that use has influenced the distribution of many species and ecosystems.

The historical range of variability (HRV) concept provides an alternative frame of reference for naturalness and gauging contemporary human influences on fire regimes. Past variations in fire frequency, magnitude, and in some cases, intensity, can be inferred in many ecosystems from analysis of tree rings, fire scars, and charcoal from lake sediments and soil. Historical variations in fire behavior in some regions are correlated with changes in climate and human activity. The relationships between HRV for fire and HRV for other environmental factors like climate on presettlement landscapes have been assumed to bracket conditions that might be considered "natural," although on many landscapes, human activities likely contributed to that variation (Landres et al. 1999, Swetnam et al. 1999, Willis and Birks 2006).

What role should these concepts play in fire management? Part of the justification for the HRV approach is that it is considered to be a conservative indicator of sustainability (Millar 1997) and provides a benchmark for restoration of perturbed ecosystems (Fulé et al. 1997). Few would question their value as benchmarks or bounds for assessing the effects of human actions and management on fire occurrence and behavior. However, HRV depends on the period on which it is based, and in most instances that period is before Euro-American interference in fire regimes. Of course the range of HRV increases as the historical timeframe increases (Millar and Wolfenden 1999).

Although significant departures from "natural" or HRV may in some cases present ecological risks, it is unclear if these concepts are appropriate as a sole basis for resource management (Vitousek et al. 2000). In some silvicultural situations and in certain applications of prescription fire to reduce forest fuels, naturalness or HRV may not be a useful reference. Fire events that might otherwise be judged as natural or within the HRV may have undesirable consequences where landscapes have been affected by human actions such as fragmentation or invasive alien species. Clearly articulating the relationship of management goals to HRV metrics, especially in an era of climate change, provides an important context for restoration. Regardless of whether HRV is used in a specific manner to set ecological restoration or management objectives, there is great inherent value in developing historical knowledge and understanding. Historical perspectives are often essential to identify dynamical behaviors, trends, and changes in ecosystems and their likely causes.

## Ponderosa Pine (Western United States)

Warnings of deleterious effects of fire suppression on semiarid forest ecosystems long preceded actions to address this issue. Cooper (1960), Weaver (1968), and Biswell et al. (1973) all showed that the historical pattern of frequent fires (one or more fires per decade from high-frequency lightning fire) in Southwest U.S. ponderosa pine forests (fig. 1a) had been disrupted by fire suppression and other land management practices. Further, they showed that reduction in burning had altered forest structure, causing accumulation of fine surface fuels, and increased density of understory saplings and smaller trees that act as "ladder fuels" that carry fire into the lower canopy (Dodge 1972).

These early reports led to a plethora of studies documenting the significant role of frequent low-intensity surface fires in ponderosa pine and other semiarid ecosystems, and documented long-term consequences of fire suppression (Allen et al. 2002, Covington and Moore 1994, Fulé et al. 2004b, Moore et al. 2004, Swetnam and Baisan 1996). Fire has essentially been eliminated for more than a century on broad portions of the forested landscape in the Southwestern United States, the result of reduction in fine grass fuels by intensive livestock grazing and effective fire suppression. The resulting accumulation of primarily woody fuels, which can intensify fire behavior and potentially carry fire into the overstory, exceeds what was present historically. Researchers have argued that these changes have resulted in increased frequency of large, high-severity crown fires in Southwest U.S. ponderosa pine forests (Allen et al. 2002, Covington and Moore 1994). Similar forest structure and fuels changes have occurred in other parts of dry, ponderosa pine-dominated forests of the inland West, such as the interior Columbia River basin (e.g., Hessburg and Agee 2003) and pine and mixed-conifer forests of the Sierra Nevada (Kilgore and Taylor 1979, Stephenson 1999, Swetnam 1993), and Colorado Front Range (Graham 2003).

## Chaparral (Pacific South Coast)

California chaparral (fig. 1b) typically burns in high-intensity crown fires, and fire spread is through shrub canopies with surface fuels accounting for little or no fire spread. Early studies characterizing differences in fire size north and south of the United States border invoked fire suppression as the primary explanation for these patterns (Minnich 1983, Minnich and Chou 1997). However, recent analyses show no evidence that $20^{th}$-century fire suppression has diminished fire activity on these landscapes (Conard and Weise 1998, Keeley et al. 1999, Weise et al. 2002). In fact, throughout the $20^{th}$ century, about a third of this region has burned every decade (Keeley et al. 1999), which reflects a relatively high fire frequency compared to

Figure 1— Representative examples of ecosystems specifically discussed in this paper. (A) Ponderosa pine forest in the Southwestern United States illustrating the open nature of surface-fire regime forests dominated by large trees with clear bole and thick bark, (B) chaparral and sage scrub shrublands juxtaposed with urban sprawl in southern California, (C) closed nature of crown-fire boreal forests with dense stocking of trees and weak pruning of lower branches, (D) Great Basin sagebrush, (E) Southern Appalachian pine and hardwood forest, and (F) Southeastern longleaf pine.

the long-term historical fire regime (Keeley and Fotheringham 2003, Minnich and Chou 1997). The fire regime in this region is dominated by human-caused ignitions, and fire suppression has played a critical role in preventing the ever-increasing anthropogenic ignitions from driving the system wildly outside the historical fire-return interval. Because the net result has been relatively little change in overall fire regimes, there has not been fuel accumulation in excess of the historical range of variability, and as a result, fuel accumulation or changes in fuel continuity do not explain wildfire patterns (Keeley et al. 2004, Moritz 2003, Moritz et al. 2004, Zedler and Seiger 2000).

High-intensity chaparral crown fires pose a major threat to economic values because urban sprawl has placed vast stretches of residential areas within a matrix of these hazardous fuels. These landscapes are vulnerable to the most costly wildfires in the United States in terms of loss of lives and property owing to the annual threat of severe fire weather fanned by autumn Santa Ana foehn winds. Since 1970, 12 of the 15 most destructive wildfires in the United States have occurred in California chaparral, costing the insurance industry $4.8 billion (Halsey 2004: 48).

The major resource threat posed by the current high-frequency fire regime is loss of native vegetation. Chaparral recovery requires two or more decades of fire-free conditions, and more frequent fires have a destabilizing effect. High fire frequency displaces native shrubs with alien annual grasses and forbs, leading to increased flammability, decreased slope stability, and loss of biodiversity (Keeley et al. 2005a). Without decreases in human ignitions, current fire suppression efforts must be sustained if we are to retain much of this ecosystem. Although fuel manipulations of ponderosa pine ecosystems may effectively reduce fire hazard on those landscapes, they are decidedly less effective on chaparral landscapes, and ultimately fire hazard reduction is likely to be achieved by directing fuel modifications away from wildland areas and more toward the wildland-urban interface. Closer integration of state and federal fire management with local land use planning would also enhance protection of urban environments and associated chaparral systems.

## Boreal Forest (Alaska and Canada)

Boreal forests (fig. 1c) are the largest biome in the Northern Hemisphere. Because of high tree density, retention of lower branches, accumulation of surface fuels, and compact arrangement of flammable fuel in the canopy, fires in North American boreal forests are dominated by crown fires with high flaming intensity and high rates of spread. These forests have a short fire season extending from June to August. Fire activity largely depends on co-occurrence of summer lightning and low fuel moisture resulting from a persistent high-pressure system (Nash and Johnson 1996).

Fire frequency has changed several times in the last 400 years more or less synchronously across the North American boreal forest, with changes apparently related to large-scale climate patterns (Bergeron and Archambault 1993, Johnson and Wowchuk 1993, Murphy et al. 2000). The hazard of burning seems to be independent of forest age, because younger and older forests have the same chance of burning, and there is little evidence that older forests have fuel accumulations that make them more flammable. Wildfires are propagated by small and medium-sized fuel, and the amounts of these fuels do not change after the closing of the forest canopy at about 20 years after the fire (Bessie and Johnson 1995, Hely et al. 2001). Of the fires that determine the age mosaic of the landscape, about 90 percent are >1000 ha and about 40 percent are >10 000 ha (Reed and McKelvey 2002). The landscape age mosaic comprises small older patches embedded within a matrix of younger forests initiated by more recent burn events. These older patches are the remnants of large burns that have been progressively reburned.

These patterns have been relatively undisturbed by humans because lightning is the dominant ignition source in most areas, and fire management has had minimal effect on most boreal forests in North America (Johnson 1992). Close to 50 percent of the area burned is the result of fires that receive no management action owing to their remote location (Stocks et al. 2003). The main exception is the southern margin of the boreal forest that has been fragmented by settlement (Mackintosh and Joerg 1935) and, particularly in the early 1900s, burned at very short intervals by escaped fires (Tchir et al. 2004, Weir and Johnson 1998). The efficacy of fuel treatments for reducing fire spread or intensity in boreal forest has not been shown.

## Great Basin Sagebrush (Intermountain West)

Much of the dryland region between the Sierra Nevada and Rocky Mountains has historically been shrublands (fig. 1d) with Great Basin sagebrush being an important dominant species (Blaisdell et al. 1982). Native understory bunchgrasses combine with forbs to form an understory with discontinuous patches between shrubs. Historical fire regimes were dominated by stand-replacement mixed surface and crown fires at variable return intervals from 35 years on moister sites to 70 to 200+ years on drier sites (Baker 2006a, Welch and Criddle 2003, Whisenant 1990). Most shrubs do not resprout and have limited seedling recruitment, and thus they gradually reestablish after fires, with full recovery of the shrub component taking from 15 to 60 years. Discontinuous fuel distribution often left unburned patches of sagebrush (Miller and Eddleman 2001), which were important parent seed sources for regeneration.

In the late 1800s, overgrazing by free-ranging cattle led to a depletion of perennial grasses and other palatable forage. The accidental introduction and rapid spread of cheatgrass in the early 1900s (Mack 1981) resulted in rapid invasion of overgrazed sagebrush rangeland (Billings 1990). As cheatgrass dominance increased, the fine fuel loads it produced added to site flammability, leading to increased fire frequency, greater continuity of fuels (which diminished unburned sagebrush seed source patches), and further decreases in native perennials, grasses, forbs, and shrubs (Knapp 1996). Adding to this problem were fire management activities such as prescription burning, introduced to increase the rangeland value of this ecosystem (Keeley 2006). Fire suppression effects have been largely eclipsed by rangeland practices that have favored the expansion of grasslands over sagebrush steppe vegetation. The destabilizing effects of grazing and fire have created systems that require assertive revegetation and strategic control of fire to reestablish species and structures that were present before the introduction of cheatgrass.

## Pine and Pine/Hardwood Forests (Southern Appalachians)

In the southern Appalachian Mountains (fig. 1e), forest composition and structure differ along gradients of topography, moisture, and elevation (Braun 1950). The role of fire across these gradients is a matter of considerable scientific debate (DeVivo 1991, Runkle 1985, van Lear and Waldrop 1989, Vose 2000) with significant implications for forest management (van Lear 1991). Moderate to high-intensity crown fires are critical for the maintenance of pine and pine/hardwood forests (dominated by pitch pine, Table Mountain pine and several species of oak in the overstory and a shrubby understory of mountain laurel and rhododendron species on dry, exposed ridges (Barden and Woods 1976, Waldrop and Brose 1999). Fire exclusion has limited the occurrence of such fires, thereby promoting increased dominance of hardwoods and a marked decline in pine populations. Selective logging in some areas has promoted establishment of dense thickets of mountain laurel, which suppressed herbaceous diversity and tree establishment, and increased the risk of intense fires (Elliott et al. 1999).

Before European settlement, oak and oak-American chestnut forests on mesic slopes were maintained by a combination of lightning and human-set fires (Abrams and Nowacki 1992, Clark and Robinson 1993). Fire suppression has been nearly 100 percent effective in these ecosystems. The elimination of fire, coupled with an array of other disturbances (e.g., logging and chestnut blight) facilitated the increased dominance of shade-tolerant species such as red maple (Abrams 1998, 2003; Crow 1988; Lorimer 1985). The role of fire in wetter areas, such as in mesic cove and northern hardwood forests, is poorly understood. It is likely that fires occurred at irregular intervals and at relatively low frequencies, probably associated with periods of extreme drought (van Lear and Waldrop 1989), and this may account for the prevalence of shade-intolerant species such as tulip poplar in some old-growth sites (Lorimer 1980).

The diversity of forest systems described above has existed as such in the southern Appalachians for only 10,000 years (Davis 1983), a period during which Native Americans actively used fire in this region (DeVivo 1991). Lightning strikes were sufficiently frequent on exposed slopes to maintain the pine and pine/hardwood forests on those sites, although human-caused ignitions were likely important across much of this forest gradient (Clark and Robinson 1993). The decline in Native American populations beginning in the 17th century may have produced significant changes in southern Appalachian fire regimes, well before modern fire suppression. Assessments of the value of fire as a management tool in this region require some consideration of the effects of burning by Native Americans on cultural landscapes.

## Longleaf Pine (Southeastern United States)

Coastal plain forests dominated by longleaf pine are among the most threatened ecosystems in the Southeastern United States (fig. 1f). In presettlement times, longleaf pine savannas occupied over 25 million ha of the Southeastern coastal plain from Texas to North Carolina; today, these forest ecosystems occupy less than 2 percent of that area, and old-growth stands account for only a few thousand hectares (Early 2004). Although much of the loss of longleaf pine savannas was caused by logging and deforestation for agriculture, historical changes in the role of fire have also played a significant role.

Longleaf pine savannas are especially well known for their high herb diversity. In moist areas that are frequently burned, herb diversity is exceptionally high and the relationship between fire and general patterns of biological diversity has been well studied (Christensen 1977, Walker and Peet 1983, Wells 1942). Many of these herbs have fire-dependent life history traits such as fire-stimulated flowering and fire-dependent patterns of growth. Exclusion of fire allows relatively few species to dominate and shade out competitors, resulting in a rapid decline in herb diversity.

As in many savanna ecosystems, frequent and low-intensity fires play a significant role in the maintenance of longleaf pine ecosystems. Presettlement fire-return intervals likely ranged between 3 and 10 years (Christensen 1981, Garren 1943, Wells 1942). Because of unique seedling characteristics, longleaf pine is especially well adapted to and dependent on this fire regime (Chapman 1932, Platt et al. 1988, Wahlenberg 1946). Disruption of historical fire regimes prevents such establishment, allows invasion of shrubs and other tree species, and creates conditions favorable to longer return intervals and higher intensity fire regimes (e.g., Myers 1985). The remnants of this ecosystem that have survived intensive land use are currently threatened by fire suppression activities.

**Scientific understanding of fire can inform policy, with the dual objectives of managing for fire-safe environments (where appropriate) and sustaining the functional integrity of fire-prone ecosystems.** The six systems discussed above illustrate regional variation in fire activity and ways in which fire management and other human activities have altered ecosystem processes. The examples present different patterns of fire hazard, fire risk (box 2), and patterns of human impact. Each system requires a different management strategy to achieve specific desired outcomes. One of the important lessons to be learned from these contrasts is that a single model of past fire regimes or appropriate fire management action is inappropriate (Johnson et al. 1998, Schoennagel et al. 2004, Veblen 2003). The diversity of North American ecosystems requires a comparable diversity in fire management, with a flexible approach that characterizes adaptive management.

# Fire Regimes as a Framework for Understanding Fire Processes

**Regionally focused fire management is premised on a consideration of spatial variation in mechanisms that drive ecosystem processes, and how these processes lead to different fire hazards in different ecosystems.** Such insights can best be gained by a clear understanding of the factors that influence fire behavior (Johnson and Miyanishi 2001), and how those factors differ across the landscape. Fire regime (Gill 1973, Heinselman 1981, Johnson and Van Wagner 1985) is an ecosystem attribute with both temporal and spatial domains (Morgan et al. 2001). Traditionally, fire regime has been defined by fire frequency, intensity, and seasonality. We suggest a more detailed definition that includes (1) fuel consumption and fire spread patterns, (2) intensity and severity, (3) frequency, (4) patch size and distribution, and (5) seasonality.

## Fuel Consumption and Fire Spread Patterns

Fires consume different fuelbed strata (sensu Sandberg et al. 2001), and each fuelbed stratum is involved in different aspects of combustion, energy release, and fire effects (Ottmar et al. 2007) (fig. 2). **Surface fires** are spread by fuels that are on the ground, which can be either living herbaceous biomass or dead leaf and stem material. **Crown fires** burn in the canopies of the dominant life forms, and the term is most usefully applied to shrub- and tree-dominated associations

*Box 2.*
## Fire Hazard vs. Fire Risk

Fire hazard refers to a fuelbed defined by volume, type, condition, arrangement, and location—these characteristics determine ease of ignition and resistance to control (National Wildfire Coordinating Group 2005). Fire hazard expresses potential fire behavior for a fuelbed, regardless of weather-influenced content of fuel moisture. Fire risk is the probability or opportunity that a fire might start, as affected by the nature and incidence of causative agents, including both natural and human ignitions. For example, data on the distribution of lightning strikes can quantify the risk of ignition for a particular landscape. Fire risk is sometimes considered to be the potential change in resource condition or value (e.g., dead trees), or change in economic value associated with human activities (e.g., structures), although these situations actually refer to values at risk.

Some examples can illustrate the contrast between fire hazard and fire risk. Temperate rain forest with a fuelbed that includes high down wood has very high fire hazard, but fire risk is very low because it is unlikely that fuel moistures will be low enough to sustain fire even if an ignition source were available. Undisturbed dense chaparral has high fire hazard because its high fuel loads can generate high fire intensities. Fire risk in this system is generally low except during the summer when fuel moistures are very low and during autumn when Santa Ana winds contribute to fire spread.

Standing dead trees in a forest that has experienced bark beetle attack have relatively low fire hazard and low risk of fires igniting and spreading through the crown. However, dead branches subsequently fall, and eventually the trees fall, contributing large amounts of fuels and increasing fire hazard over time.

Fuel reduction projects in forests are intended to reduce fire hazard by reducing surface fuels, continuity of fuels from the surface to the canopy, and continuity of fuels within the canopy. Fire risk is unaffected, but if a fire does occur, fire intensity and effects on the overstory may be less owing to the lower fuel loading. Fuel reduction projects near roads may have the unintended consequence of increasing annual weeds that generate highly combustible fuels (increased hazard), and thus facilitate ignitions (increased risk).

The relative effects of hazard versus risk differ across ecosystems and fire regimes. For example, high fire risk is normal in ponderosa pine forests that are resilient to frequent fire, but high fire hazard, which may occur following many decades of fire exclusion, can damage the overstory if fuel loadings are high enough to cause high flame lengths. In contrast, sustainability of chaparral shrublands is threatened when fire-return intervals are long, because high fire intensity does not typically affect recovery and sustainability of this ecosystem.

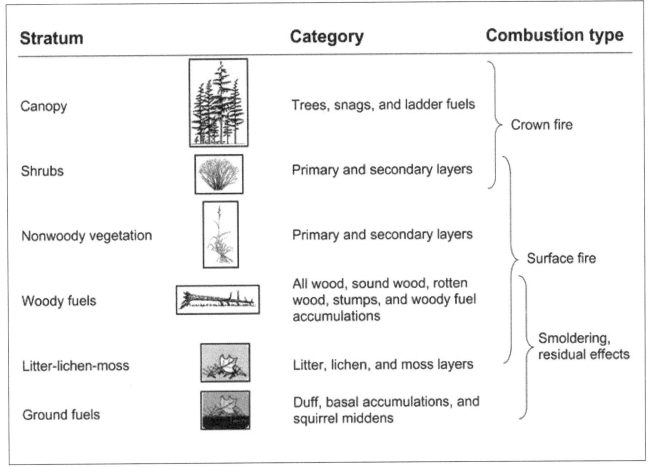

Figure 2—Fuelbed strata and their involvement in different types of combustion (from Ottmar et al. 2007).

(Scott and Reinhardt 2001, Van Wagner 1977). Crown fires tend to be less common in hardwood forests because of greater foliage moisture and lower canopy bulk density. **Passive crown fires** spread in surface fuels and then are carried into the canopy by shorter ladder fuels, often called "torching." **Active crown fires** are spread by both surface fuels and canopy fuels, but **independent crown fires** are not linked to surface fires, and generally require rather dense canopies and sufficient wind or steep terrain to carry fire. All of these surface and crown fire types are characterized by flaming combustion, whereas **ground fires** spread slowly by smoldering combustion through duff (or peat) and can be sustained at relatively high fuel moisture conditions (Miyanishi 2001). Because they can smolder for long periods, perhaps months, they may "store" ignitions from lightning fires during times when weather conditions are less suitable for more active burning, and then erupt into surface or crown fires with changes in weather or fuels.

Surface fires and crown fires have different effects on ecosystem processes and on the evolution of plant traits. For example, thick bark and self-pruning of lower branches are common traits in pines dominant under surface fire regimes, but thin bark, weak pruning, and serotinous cones are traits restricted to crown fire ecosystems (Keeley and Zedler 1998).

Some ecosystems are characterized by either surface fires or crown fires, but in many systems, mixtures of both fire types are common. These are sometimes called "mixed fire regimes" typified by a combination of surface fires and passive crown fires. The proportion of landscapes burned in one or the other fire type is a function of the time since last burning, rate of fuel accumulation, antecedent drought, and severity of fire weather. Sometimes such fires are referred to as being of moderate fire severity, but they are more properly called mixed-severity fires. Besides such spatial mixtures, some ecosystems experience a temporal mix of surface fires alternating in time with high-severity crown fires (Zimmerman and Omi 1998).

## Fire Intensity and Severity

Multiple burning patterns can occur in any given fire (fig. 3), with variation typically expressed by the terms intensity and severity. Fire intensity refers to the rate of energy release, or to other direct measures of fire behavior such as temperature or flame length. Fire severity refers to injury, loss of biomass, or mortality resulting from fire (Moreno and Oechel 1994). Although fire intensity and fire severity are often correlated, this is not always so. For example, high tree mortality commonly results from fires that burn actively in the canopy; however, fires that smolder in the duff are also lethal to some plants and animals (Sackett et al. 1996). Winter prescription burns in California chaparral typically generate lower fire intensities, but are more lethal to shrub regeneration (Keeley and Fotheringham 2003).

For many purposes the best physical descriptor of fire intensity is fireline intensity, which is the rate of heat transfer per unit length of the fire line (kW/m) (Byram 1959). This represents the radiant energy release in the flaming front, but is not specifically a measure of temperature (Alexander 1982). This is an important characteristic for propagation of a fire and thus has been built into models of fire behavior used during fire suppression activities in the United States (Rothermel 1983). In practice, flame length has been found to correlate with fireline intensity and is often used in such models because it is easier to measure (Andrews 1986). However, this relationship has not been widely tested, and accuracy likely differs depending on the ecosystem (Cheney 1990).

Figure 3—The Aspen Fire burned over about 34 000 ha in June 2003 in the Santa Catalina Mountains near Tucson, Arizona. This human-ignited fire burned in a mosaic pattern of mixed severity, with (foreground) understory surface burn, including small patches of passive crown fire, and (background) active crown fire in ponderosa pine and mixed conifer on the steep slopes. Over 200 homes and commercial buildings burned in the village of Summerhaven, located just below the mountain ridgeline at right center in the photograph.

Fireline intensity has been used to predict scorch height of conifer crowns and other biological impacts of fire (Albini 1976, Borchert and Odion 1995), whereas other system components such as non-wettable layers in soil may be more closely related to duration of soil heating (DeBano 2000), and survival of seed banks may be more closely tied to maximum soil temperatures (Bradstock and Auld 1995).

Despite the importance of fire-intensity measures, fire managers do not always have the luxury of controlled experiments and are faced with describing fires after they have occurred. Fire effects such as extent of biomass loss and mortality are termed fire severity, and these are often correlated with fire intensity (e.g., Dickinson and Johnson 2001, Moreno and Oechel 1994). In ecosystems characterized by crown fire, all aboveground biomass is typically killed, and thus in these systems mortality may not be strongly tied to fire intensity. Fire intensity can have an effect on postfire resprouting of hardwoods and shrubs and thus is sometimes considered a measure of fire severity. However, because some species are incapable of resprouting, this cannot be used as a measure of fire severity without accounting for spatial variation in community composition.

Fire severity is often interpreted as a measure of ecosystem effects, defined as the capacity for regeneration of plant cover and community composition as well as recovery of hydrologic processes (National Wildfire Coordinating Group 2006). However, fire severity and ecosystem responses should be considered separately. Although they may be closely coupled in some ecosystems (e.g., in some forest types, high fire severity is coupled with poor regeneration), they are largely uncoupled in other ecosystems (e.g., in California chaparral, high fire severity is only weakly tied to the capacity for revegetation) (Keeley et al. 2005a). Also, watershed hydrologists often describe fire severity in terms of damage to physical soil structure that may affect erosion processes (Moody and Martin 2001), but although fire per se consistently affects watershed hydrology, the degree of fire severity sometimes does not (Doerr et al. 2006).

> Assessing fire frequency can involve considering complex fire behavior at different spatial scales.

## Fire Frequency

**Fire frequency** is the number of occurrences of fire within an area and time period of interest. **Fire rotation interval** is the time required to burn the equivalent of a specified area, whereas **fire return interval** is the spatially explicit time between fires in a specified area. For example, wildlands in southern California have an average fire rotation interval of 36 years, but this can range from fires every few years at some sites to fires every 100 years at other sites (Keeley et al. 1999).

Assessing fire frequency can involve considering complex fire behavior at different spatial scales. At very broad spatial scales, fire frequency in ecosystems characterized by crown fire, such as the boreal forest or sagebrush, involves stand replacement and is documented in fire atlases or by time-since-last-fire (stand age) maps estimated from aerial photography and tree rings (fig. 4). One limitation to determining the historical extent of crown fires in forests is that many of the lower elevation forests of western North America have been logged, making it difficult to determine if large fires ever occurred on much of this landscape.

In surface-fire regimes, low-intensity fires documented in fire-scarred trees provide a unique record of long fire histories that typically span 200 to 500 years (fig. 5), and in the case of giant sequoias about 3,000 years (Swetnam 1993). Tree-ring-dated fire scar records have temporal resolutions of years

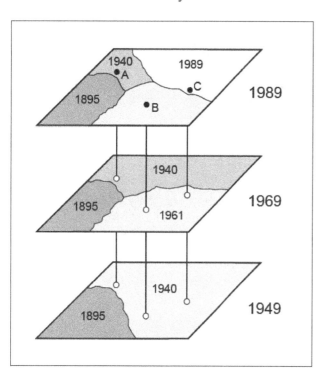

Figure 4—Layers making up time-since-last fire map created by burning over of previous burns.

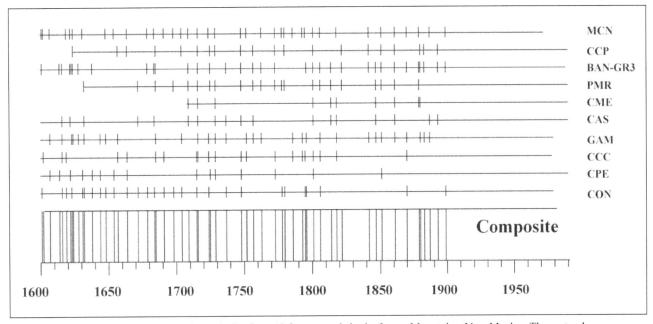

Figure 5—A 400-year set of fire-scar chronologies from 10 forest stands in the Jemez Mountains, New Mexico. These stands are broadly distributed around the mountain range, spanning an area of about 50 000 ha. The horizontal lines and tick marks in the upper graph show timespans and fire dates, respectively, of fires recorded by any sampled fire-scarred tree within each of the stands. The bottom graph shows the same stand chronologies, but only fire dates recorded by 25 percent or more of the trees within each of the stands. The long vertical lines at the bottom show the composite of all dates for each graph. Note that the 25 percent filter emphasizes fires that were probably relatively widespread within and among stands. The surface-fire regime disruption ca. 1900 is evident in both graphs. Early and persistent disruption of the fire regime is evident in the three lowermost stands (CCC, CPE, and CON); this is attributed to early livestock grazing in these specific sites. An early-1800s gap in fire occurrence in all chronologies is most apparent in the 25 percent filtered chronologies (bottom graph), possibly caused by a decadal-scale cool period during this interval (Kitzberger et al. 2001). MCN = Monument Canyon Natural, CCP = Capulin Canyon, BAN-GR3 = Ban-Group 3 (Apache Mesa), PMR = Pajarito Mountain Ridge, CME = Camp May East, CAS = Cañada Bonito South, GAM = Gallina Mesa, CCC = Clear Creek Campground, CPE = Cerra Pedernal, CON = Continental Divide.

and seasons (Dieterich and Swetnam 1984), which enable detailed spatial analyses when sampled over defined areas (e.g., Reed and Johnson 2004). Fire-scar dendrochronology has shown that fire frequency differs in a fine-grained spatial pattern, often with marked differences between north- versus south-facing slopes or upper slopes versus lower slopes (Caprio 2004, Caprio and Graber 2000, Hessl et al. 2004, Norman and Taylor 2002). In addition, regional networks of fire scar chronologies often show synchronous fire events among multiple watersheds and mountain ranges, and these events are often well-correlated with drought and atmospheric circulation indices (e.g., Hessl et al. 2004; Kitzberger et al. 2001, 2007; Swetnam and Betancourt 1990).

Charcoal and pollen deposits can provide fire frequency estimates covering the past 10,000 years or longer, but typically at temporal resolutions of decades to centuries (Clark and Robinson 1993, Millspaugh et al. 2004). These studies have shown vegetation changes in concert with changes in climate and fire (Whitlock et al. 2004). Of particular importance is the recognition that fire regimes have differed markedly throughout the Holocene such that fire regimes present at the time of Euro-American contact were in some cases relatively short-lived phenomena that were preceded by different fire regimes in earlier times (Millspaugh et al. 2004).

Each of these fire-dating approaches presents challenges to correctly interpreting fire occurrence measures. Fire-scar records from individual trees can approximate the frequency of fire that occurred around a particular tree, but because a minority of trees in forests are scarred in surface fire regimes, and not all previously fire-scarred trees are rescarred during subsequent fires, fire event records from single trees are generally considered conservative estimates of point-fire occurrence. A composite fire frequency can be generated for a forest stand on the scale of about 10 ha or larger, with standwide fire frequencies estimated by an inventory of fires that scarred some minimal percentage (e.g., 25 percent) of sampled fire-scarred trees during the same year within the stand (Dieterich 1980, Swetnam and Baisan 1996). At the stand scale, this method captures the frequency of relatively widespread fire events (if samples are well distributed) but ignores intrastand spatial variation (e.g., fig. 6). Thus, for a given point, it is potentially an overestimate of fire frequency. Fire frequencies from fire-scar composites (or any other reconstruction method) differ as a function of the study area and sample size (Baker and Ehle 2001, Falk and Swetnam 2003, Hessl et al. 2004, Van Horne and Fúle 2006, Veblen 2003). Fire history reconstructions based on stand age and structure (e.g., Johnson and Gutsell 1994) are limited by low spatial resolution of past fire perimeters and intrastand variations, low temporal resolution of some past fire dates, and potential biases in model estimations of stand-age distributions and fire frequencies (Finney 1995).

Fire frequency estimates based on charcoal deposition are affected by wind patterns that affect dispersion of particles, which in turn are affected by particle size, which in turn is a function of fuel type, as well as sediment movement. Charcoal abundances in sediment cores may be functions of both fire frequency and severity, with concentrated charcoal layers (or charcoal "peaks") in the time series reflecting either individual high-severity events, frequent fire periods, concentrated erosion periods, or all of these processes in combination. Documentary sources of fire history (e.g., repeat aerial photographs and fire atlases) are also subject to errors, omissions, and problems of low temporal or spatial resolution (Morgan et al. 2001). Despite these limitations of data and methods of fire history reconstruction, both paleoecological and documentary sources have proven useful in providing knowledge of past fire regimes and their controls across a broad range of spatial and temporal scales (Morgan et al. 2001).

## Fire Patch Size and Distribution

Fire size differs from a lightning-ignited fire that remains localized around the tree it strikes, to massive boreal forest crown fires that burn millions of hectares. On most landscapes, a small proportion (5 percent or less) of fires account for 95 percent of the area burned (Strauss et al. 1989). This means that it is primarily the very large fires in the tail of the size distribution that determine the age distribution and spatial age mosaic of the landscape. Thus, for both ecological and practical reasons, large fires are often of most concern to fire and resource managers.

Distributions of overall fire size differ regionally and between surface fire and crown fire regimes. Likewise the size of different fire-severity patches within fire perimeters may differ greatly, creating a mosaic of patches (fig. 6). Many forests exhibit complicated patterns of fuel consumption, comprising a mixture of surface

**Many forests exhibit complicated patterns of fuel consumption, comprising a mixture of surface fire, crown fire, and unburned patches.**

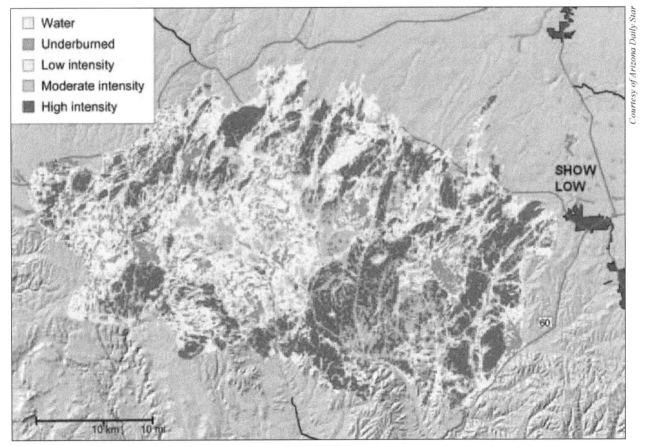

Figure 6—Mosaic fire pattern mapped for the Rodeo-Chedeski Fire, Arizona. Mapping was done by a Burned Area Emergency Response team, using a combination of remotely sensed data and on-the-ground observations. Severity categories were largely qualitative and coarse resolution, intended mainly for showing approximate spatial patterns of burn severity. High severity indicates locations where all or most vegetation was blackened and killed, and the ground was covered only with ash. Moderate severity indicates a mosaic of green areas and burnt areas, and the ground was covered with a mixture of ash, leaf litter, and unburned organic matter. Low severity indicates that some trees, shrubs, and grasses were burnt, but most of the vegetation remained green.

fire, crown fire, and unburned patches. This heterogeneity is important to ecosystem processes such as tree recruitment (Bonnet et al. 2005). For example, in the mixed-conifer forests of the Sierra Nevada in California, patches of high-intensity fires produce light gaps for tree regeneration (Rocca 2004, Stephenson et al. 1991). These gaps also accumulate fuels at a slower rate, and thus have a greater probability of fires missing them until saplings reach sufficient size to withstand fires (Keeley and Stephenson 2000).

Patch distribution at large spatial scales differs among ecosystems and affects patterns of vegetation recovery and habitat structure for animals. Mixed-conifer forests in the Western United States are particularly sensitive to patch-size distribution. The historical fire regime was often a mixture of surface fires, which left dominant trees alive, and passive crown fires that killed all trees within small patches from a few hundred square meters to a few hectares. A similar pattern may have prevailed in ponderosa pine forests in the central and northern Front Range of Colorado (Brown et al. 1999, Ehle and Baker 2003, Sherriff 2004). When patch size is hundreds or thousands of hectares, regeneration may be limited because the dominant trees lack a dormant seed bank, either in the soil or stored in serotinous cones. Reproduction (at least in the short term) requires mixed fire regimes that generate gaps in the canopy but allow survival of parent seed trees within dispersal distance (Allen et al. 2002, Greene and Johnson 2000, Savage and Mast 2005, Weyenberg et al. 2004). In boreal forests, the area of unburned patches per unit of area burned may remain constant during periods when climate is not greatly changing (Johnson et al. 2003). Thus, despite variation in fire size (taken to be the total area within the burn perimeter), the maximum dispersal distance either from the burn perimeter or from surviving patches typically is not greater than about 150 m (Greene and Johnson 2000).

Chaparral shrublands commonly experience large crown fires that cover significantly more than 10 000 ha. Heterogeneity of fire severity patches within the overall perimeter is relatively low as fires burn in a rather coarse-grained pattern of uniform high severity. This poses no threat to most plant species in these systems because regeneration mostly depends on dormant seed banks and resprouting from basal lignotubers. However, such large fires may inhibit recovery of large fauna that must disperse back into burned areas, a management concern in chaparral landscapes fragmented by roads and structures.

## Fire Seasonality

Fire seasonality is a function of the coincidence of ignitions with fuel conditions. Fire seasons generally center around the driest time of the year, but other factors may be involved. For example, in monsoon climates of the Southwestern United States, most area burned occurs in May or early June, whereas most fires are ignited in late June or early July when monsoon lightning storms break a several-month spring drought. Fires in eastern deciduous forests tend to be concentrated in late winter and early spring, coincident with surface leaf litter accumulation dried by open canopies. Mediterranean climates have fires spread out through the summer until ended by winter rains. In southern California, fire season may last 6 to 9 months, whereas in boreal forests, it may be constricted to 1 to 3 months, depending on annual climate patterns.

The peak numbers of ignitions do not always coincide with peak area burned, particularly where human-caused ignitions dominate. Season of burning affects types of fuels consumed, fire intensity, and composition of postfire herbaceous vegetation (Knapp et al. 2005, Snyder 1986). In California chaparral, winter burning may limit postfire recovery because of the truncated winter-spring growing season for postfire vegetation (Keeley 2006).

## Climate and Weather Effects on Fire Regimes

Climate and weather affect fire regimes in a diversity of ways in North American ecosystems, and understanding these relationships will improve predictions and management of future fire activity. Climate comprises atmospheric processes that characterize broad spatial and temporal scales ($10^4$ to $10^9$ $km^2$, seasons to millennia), whereas weather encompasses relatively fine-scale processes (1 to $10^4$ ha, minutes to seasons). Recent advances in fire climatology have led to the development of long-range fire forecasting tools that are most appropriate for regional scales and seasonal planning. Approaches include statistical associations between seasonal and interannual climate with regional fire activity (Collins et al. 2006, Westerling et al. 2002) and use of mechanistic models to predict fire responses to climate changes (e.g., Flannigan et al. 2000, Lenihan et al. 2003). Fire meteorology focuses on the fine-scale weather and other physical processes that drive fire behavior, and are used both in firefighting operations and to differentiate the relative roles of weather and fuels in determining fire behavior. The influence of weather conditions on fire behavior has been incorporated into fire behavior and spread models and fire danger rating systems (e.g., Finney 1998, Rothermel 1983, Van Wagner 1987).

## Climate and Fire Activity

Climate affects fire regimes by affecting fuel moisture, and thus flammability, and by changing patterns of primary productivity, and thus fuel quantity. Climate, of course, also affects the frequencies and magnitudes of various weather variables occurring at finer temporal and spatial scales. Over much of the Western United States there is a strong seasonal to interannual link between precipitation and fire with various time lags (Gedalof et al. 2005, Westerling et al. 2002). The negative correlation between fire activity and current-year rainfall is a direct consequence of effects on fuel moisture. However, 1- to 2-year lags with a positive relationship between rainfall and fire activity may reflect the effects of moisture on herbaceous fuel loads (Donnegan et al. 2001, Grissino-Mayer and Swetnam 2000, Knapp 1998, Westerling et al. 2002). Support for this interpretation comes from the lack of such lags in vegetation types without a substantial herbaceous fuel component (Littell 2006), such as in some Southwestern U.S. and Sierra Nevada mixed-conifer forests (Swetnam and Baisan 1996, 2003) and southern California chaparral (Keeley 2004).

**Climatic variability over the last century may have had a greater role than management activities in changes in fire behavior and effects in some regions and ecosystems.**

Climatic variability over the last century may have had a greater role than management activities in changes in fire behavior and effects in some regions and ecosystems. Recent studies show correlations among warming temperatures, earlier springs, and increased numbers of large forest fires in some parts of the Western United States (Westerling et al. 2006), and in Canada (Gillett et al. 2004). Anticipated warming trends as a consequence of greenhouse gas accumulation may lead to further increases in the numbers of large fires and total area burned in some regions (Brown et al. 2004, Flannigan et al. 2005, McKenzie et al. 2004). However, global climate changes are expected to produce large changes in vegetation distributions at unprecedented rates, particularly in semiarid fire prone ecosystems (Allen and Breshears 1998). These anticipated changes in fuel distribution could reduce fire activity in some regions and lead to unanticipated impacts on future fire regimes.

Climate signals are likely responsible for regional synchrony in fire activity evident in many parts of the Western United States (e.g., Swetnam and Baisan 2003, Weisberg and Swanson 2003). Similar relationships are evident in earlier warmer periods such as the Medieval Warm Period (1000 to 650 years B.P.) that have been shown to be associated with increased fire frequency (Clark 1988, Swetnam 1993, Umbanhowar 2004), and incidence of large fires (Millspaugh et al. 2004) in some regions. Climate-controlled changes in fuel production may also explain longer term patterns in fire activity. Higher levels of biomass may reflect the shift from cooler and drier conditions of the Little Ice Age (500 to 100 years B.P.) to warmer moister conditions of the $20^{th}$ century (Mann et al. 1998), which may be partially attributable to human-caused forcing (Meehl et al. 2003). The climatic and ecological effects and timing of the Medieval Warm Period and Little Ice Age were highly variable (Hughes and Diaz 1994); some regions show no evidence of one or both of the episodes, and where they did occur, the timing of the warmest or coldest phases are sometimes asynchronous between regions. Therefore, without independent historical climate evidence, it cannot be assumed that the predominant conditions of these periods occurred everywhere.

Climate and weather control fire behavior ultimately through their effect on fuels. Fuels must be dry enough for fires to be propagated; the drier the dead fine fuels, the more fuel is involved in combustion and the more heat can be produced to drive moisture from live fuels. Fuels dry only when the weather is warm and dry, and that occurs when persistent high pressure systems block the normal westerly progression of highs and lows in the Northern Hemisphere. Thus, large fires are primarily controlled by large-scale mid-tropospheric anomalous patterns that affect the synoptic-scale weather and the amount of surface heating (Bessie and Johnson 1995, Gedalof et al. 2005, Schroeder et al. 1964).

Several climate patterns produce such blocking high-pressure systems in parts of North America and create extreme fire weather. The El Niño-Southern Oscillation (ENSO), with the alternating El Niño (warm phase) and La Niña (cool phase) events, is manifested as sea surface temperature anomalies in the tropical Pacific Ocean and associated changes in atmospheric pressure and circulation patterns. El Niño is linked to wetter winter and spring conditions and reduced area burned in the Southeastern and Southwestern United States (Beckage et al. 2003, Simard et al. 1985, Swetnam and Betancourt 1990, Veblen et al. 2000). This pattern is typically reversed in the Pacific Northwest and Central and South America, where El Niño events are often associated with drier conditions and increased fire occurrence (Hessl et al. 2004, Heyerdahl et al. 2002, Kitzberger et al. 2001). La Niña typically produces the reverse pattern, with severe winter-spring droughts and large fires in the Southwest, and reduced fire activity in the Northwest (Kitzberger et al. 2007, Schoennagle et al. 2005). These are general patterns, and ENSO events vary in strength and effects on climates and fire occurrence in particular regions.

The Pacific North America (PNA) pattern and the associated Pacific Decadal Oscilllation (PDO) affect area burned in the northwestern and interior Western United States and Western Canada (Johnson and Wowchuk 1993, Skinner et al. 1999). The positive mode of the PNA is characterized by an anomalous strong trough of low pressure over the North Pacific, upstream of a ridge of high pressure over western and eastern North America. The location of the high generally extends from the Canadian Rocky Mountains in Alberta to the interior Western

United States. When such conditions occur in spring or summer, the blocking high produces an extended period of warm, dry weather that causes extreme drying of forest fuels. This pattern has been associated with most of the big fire years in the past 20 years in the Southern Canadian Rocky Mountains and interior Western United States.

The frequency of these large-scale atmospheric patterns and their associated blocking highs, particularly in spring and summer, largely determine the frequency of severe fire weather and likelihood of high-intensity fires that burn large areas. Historical variability in these synoptic conditions makes it difficult to infer the relative influence of climate and management activities (e.g., fire suppression that leads to fuel accumulation) on fire activity. Even in relatively recent times, climate shifts could have affected fire activity. For example, since the 1970s the PNA (and PDO) pattern has changed, resulting in a deeper Aleutian low shifted eastward (Trenberth and Hurrell 1994), accompanied by increases in sea surface temperatures along the west coast of North America (Hurrell 1996). Besides ENSO, PDO, and PNA climate-fire associations, some recent studies of modern and paleo records (fire scars) have identified multidecadal correlations of the Atlantic Multidecadal Oscillation (AMO) and fire occurrence time series from western North America (Brown 2006, Collins et al. 2006, Kitzberger et al. 2007, Sibold and Veblen 2006).

The oscillatory climate patterns mentioned above reflect a revolution in our understanding of the ocean-atmosphere system, with implications for fire climatology and the biogeography of fire. These climate-fire patterns are more-or-less persistent over periods of seasons to decades, and are "quasi-periodic" (i.e., not classically cyclical, but recurrent within a particular range of periods). The temporal persistence and quasiperiodic nature of these events and processes mean that long-range fire danger can potentially be forecast as an aid to fire managers and planners.

## Fire Weather

Weather conditions sufficient to allow combustion and fire spread differ among fire regimes. For example, surface fires typically burn dead biomass, and the threshold for fire spread occurs at lower windspeeds and higher relative humidities than for crown fires in which fuels are commonly living material (Weise et al. 2003). Large fires that resist suppression efforts occur under severe fire weather conditions that include high temperatures, low humidities, and high surface winds (Brotak and Reifsnyder 1977, Schroeder et al. 1964). The largest fires generally are associated with the extremes of these conditions, as illustrated by the Hayman Fire in Colorado (June 2002). The previous 2 years were warm and dry, which promoted drying of fuels. During the first 10 hours, the fire consumed less than 500 ha, but after a shift in weather that brought wind gusts up to 85 km per hour, with 5 to 8 percent relative humidity, the fire consumed nearly 25 000 ha in the subsequent 24 hours (Graham 2003).

Synoptic or regional weather patterns that generate high winds are a major determinant of fire size on some landscapes. Wind increases combustion by mixing of oxygen within fuelbeds and by altering the flame angle such that there is greater heating of fuels ahead of the flaming front. Lacking significant wind, fires develop plumes that convect heat vertically and do not preheat fuels ahead of the flaming front (Rothermel 1991). Thus, it is to be expected that fuel treatments such as understory thinning would be less effective as windspeed increases.

In the eastern half of the United States, large fires appear to be associated with intense high-pressure troughs that bring strong winds without surface precipitation during the passage of a cold front (Brotak and Reifsnyder 1977). Foehn winds (strong warm dry winds that move down the lee sides of mountains) are also often associated with large uncontrollable fires in some mountainous regions. For example, in southern California, Santa Ana winds occur when a high-pressure system centered over the Great Basin coincides with a low-pressure trough off the California coast (Schroeder et al. 1964), reversing the normal pressure gradient that causes onshore breezes from the Pacific Ocean. The air is channeled south and west out of the Great Basin around the northern and southern end of the Sierra Nevada. These dry, gusty continental winds lose their moisture on the windward ascent and are further dried through adiabatic warming on the leeward descent. They not only cause excessive drying of fuels but can turn wildfires into firestorms. These winds typically occur in fall and early winter, after the summer dry season in southern California and are associated with most large fires in the region. As human populations have increased in this area, ignitions during severe weather events have also increased (Keeley and Fotheringham 2003).

Model studies also conclude that fire spread and intensity are more sensitive to weather variables than to fuel (Bessie and Johnson 1995). Comparative study of five different fire models that were designed for landscapes as diverse as Australian eucalyptus forests and northern Rocky Mountain conifer forests, all with mixed-severity or crown fire regimes, consistently showed a strong connection between weather, climate, and fire, and a lesser role for fuels (Cary et al. 2005).

It has been argued that, historically, fires in some vegetation types such as ponderosa pine savanna were not controlled by fire weather, and contemporary weather-driven high-severity fires in these forests are related to fire suppression and elevated fuel accumulations (Agee 1997). Fuel accumulation and forest structure changes are likely involved in recent fire regime changes in Southwestern U.S. ponderosa pine and mixed-conifer forests (e.g., Allen et. al. 2002, Fulé et al. 2004a) and in parts of the interior Pacific Northwest (Hessburg and Agee 2003), although crown fires of some unknown spatial extent may have played a natural role in these forest types in other regions (Ehle and Baker 2003, Pierce et al. 2004, Sherriff 2004). On some North American landscapes, weather effects on fire behavior are far more critical than antecedent climate impacts on fuels. For example, predictable annual autumn foehn winds in southern California are the primary determinant of large fires (Schroeder et al. 1964), and therefore droughts show little or no relation to interannual variation in area burned (Keeley 2004). However, droughts are associated with a lengthening of the fire season outside the foehn wind season.

## Biogeographical Patterns of Fire Regimes

Fire regime parameters differ in space and time and are affected by a complex set of factors. Nevertheless, there are patterns that can be recognized and used in designing fire management strategies. Fuel consumption forms the basis of most classification schemes, the most basic scheme being surface fire regimes, crown fire regimes, and mixed surface and crown fire regimes. These patterns are linked to differences in fire frequency and fire intensity such that modal groupings that capture much of the landscape variation in fire regime parameters can be recognized. For most applications, fire regimes can be categorized into three general classes of intensity and frequency: low-intensity, high-frequency surface fire; high-intensity, low-frequency crown fire; and mixed-severity fire regimes.

Schmidt et al. (2002) partitioned surface fire regimes into those in which surface fire burns under a canopy of overstory trees and those that burn in the open. They partitioned crown fire regimes into those that burn at frequencies of a century or less and those that burn very infrequently (table 1). They also classified contemporary landscapes based on departure from historical fire occurrence (table 2). These classes represent modal points on a continuum of fire regimes, and fire regimes in a particular vegetation type may differ regionally. For example, ponderosa pine forests in the Southwest generally burned frequently at low intensities, but farther north in the Rocky Mountains, some ponderosa pine had a mixed-severity fire regime (Schoennagel et al. 2004, Veblen et al. 2000).

Although this simple classification explains much of the variability among ecosystems, the multiple factors discussed earlier combine to create a wide variety of multidimensional fire regimes. Patterns of ignition and timing of burning differ regionally and in concert with seasonal changes in climate (Bartlein et al. 2003). In

Table 1—Fire regime types[a][b]

| Fire | Fire-return interval | Fire spread driven by | Fire intensity | Fire effects | Ecosystem examples |
|---|---|---|---|---|---|
| | *Years* | | | | |
| I | 1–35 | Surface and other low understory fuels | Heavy understory and fuel consumption | Low to moderate fuel overstory mortality | Ponderosa pine, longleaf, pine oak savanna |
| II | 1–35 | Mostly surface fuels | Low to moderate | Aboveground biomass killed, most fuels consumed | Grassland, low scrub |
| III | 35–100 | Surface and canopy fuels | Mixed high and low | High understory mortality and fuel consumption, thinning of overstory | Western mixed-conifer, forest Appalachian pine-hardwoods |
| IV | 35–100 | Mostly canopy fuels | High | Aboveground biomass killed, high fuel consumption | Chaparral, boreal forest, sagebrush |
| V | >200 | Mostly canopy fuels | High | Aboveground biomass killed, high fuel consumption | Lodgepole pine forest, subalpine forest, Eastern U.S. deciduous forest |

[a] These are modal groups from a continuum of patterns seen in nature. See Kilgore (1987) for summary review of fire regime literature.
[b] Source: Modified from Schmidt et al. 2002.

Table 2—Fire condition classes categorizing potential vegetation on landscapes for departure from historical fire regimes[a]

| Condition class | Risk of ecosystem change | Condition of contemporary fire regimes |
|---|---|---|
| 1 | Low | Falling well within the historical range of variability |
| 2 | Moderate | |
| 2a | | Fire frequency at the low end of the range |
| 2b | | Fire frequency at the high end of the range |
| 3 | High | |
| 3a | | Fires excluded to the point where multiple expected fire-return intervals have been missed |
| 3b | | Fires greatly increased to the point where resilience thresholds are exceeded and type conversion occurs |

[a] Source: Modified from Schmidt et al. 2002.

southern California (fig. 7b) and the eastern Appalachians (fig. 7e) human-caused ignitions dominate, but with different seasonal patterns. There is substantial regional climate variation that exhibits different patterns even within similar fire regime types. For example, peak burning in longleaf pine (fig. 7f) coincides with peak lightning fires in July, whereas the same fire regime in ponderosa pine (fig. 7a) exhibits peak burning earlier in the season and offset from the lightning fire peak. Crown fire regimes in the boreal forest (fig. 7c) exhibit a June peak in burning that is driven largely by lightning, whereas southern California chaparral (fig. 7b) has an autumn peak, and lightning plays only a minor role.

**There is substantial regional climate variation that exhibits different patterns even within similar fire regime types.**

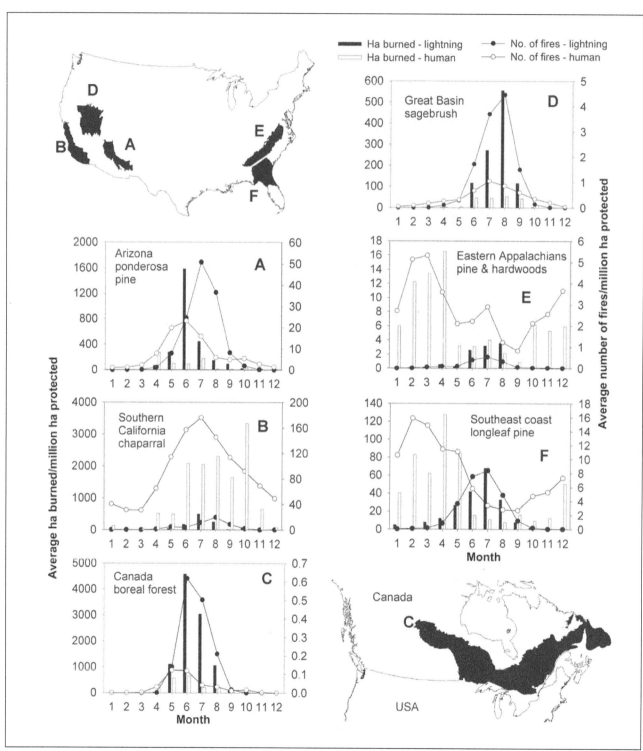

Figure 7—Seasonal distribution of lightning-ignited and human-ignited fires and area burned per million ha protected for (A) central Arizona ponderosa pine dominated landscape, (B) southern California coastal chaparral, (C) Canada boreal forest, (D) Great Basin sagebrush, (E) southeastern Appalachian pine and hardwood, (F) Southeast Coastal Plain longleaf pine landscape. A, B, D, E, and F are based on data from Schmidt et al. 2002, (A) subregions 54 and 59; (B) Santa Barbara, Ventura, Los Angeles, San Bernardino, Riverside, Orange, and San Diego Counties, (D) subregion 12, (E) subregions 43 and 59, (F) subregion 55. C is based on the Canada Large Fire database, Canadian Forest Service, Boreal Shield Ecozone, fires >200 ha for 1949 to 1999.

## Recent Changes in Fire Regimes

Detecting trends is complicated by the fact that during the 20$^{th}$ century, there has been considerable annual variation in area burned relative to area protected (fig. 8). The highest year of burning has occurred within the last two decades in the Southwest (fig. 8a), southern California (fig. 8b), the Great Basin (fig. 8d), and Canada (fig. 8c), making this period stand out, not only in these figures but in the minds of the public as well. In addition, in some of these regions, the frequency of high fire activity years has been greater in recent decades.

In the Southwest, one or more fires (or fires that joined to form complexes) exceeded 20 000 ha in every year between 2000 and 2004. Before this period, fires of such magnitude were uncommon. Several fires exceeding 40 000 ha occurred in 2003 and 2004. The 168 000 ha Rodeo-Chediski Fire (central Arizona, 2002) was two fires that merged, and collectively this event was many times larger than any single fire in Southwestern conifer forests during the previous century.

The historical record for Canada illustrates a pronounced recent change in fire activity (fig. 8c). Some have questioned whether or not this is driven by artifacts of sampling such as changes over time in area protected (Murphy et al. 2000), because for most regions, the size of the sample area from which fire statistics are drawn tends to increase with time (Podur et al. 2002). Van Wagner (1987) addressed this issue by incorporating a correction factor to account for historical changes in area sampled, and this correction is incorporated into the Canadian Large Fire database on which fig. 8c is based. However, this correction does not appear to account for all of the areas Stocks et al. (2003) indicated were likely missing from the early records. Other measures of fire activity, though, suggest that the recent increase in fire activity in the last two decades is not an artifact of sampling different size areas (Girardin 2007).

Such increases in recent fire activity are not characteristic of all regions. Indeed, in the Southeast (fig. 8e) fire activity has declined in recent decades. In southern California (fig. 8b), high fire activity years have occurred at periodic intervals throughout the 20$^{th}$ century, and there are no obvious trends in area burned. The magnitude of area burned (fig. 8) shows that, for most decades throughout the 20$^{th}$ century, southern California has had a substantially greater proportion of its landscape burned than any other region considered here.

Although recent area burned in the Southwest was exceptional on the scale of the past century, longer historical records estimated from newspaper accounts indicate that some 19$^{th}$-century fires in the Southwest exceeded 400 000 ha (Bahre 1985). Broadly synchronous 17$^{th}$- to 19$^{th}$-century fire-scar dates recorded across many Southwestern mountain landscapes lead to similar conclusions: much larger

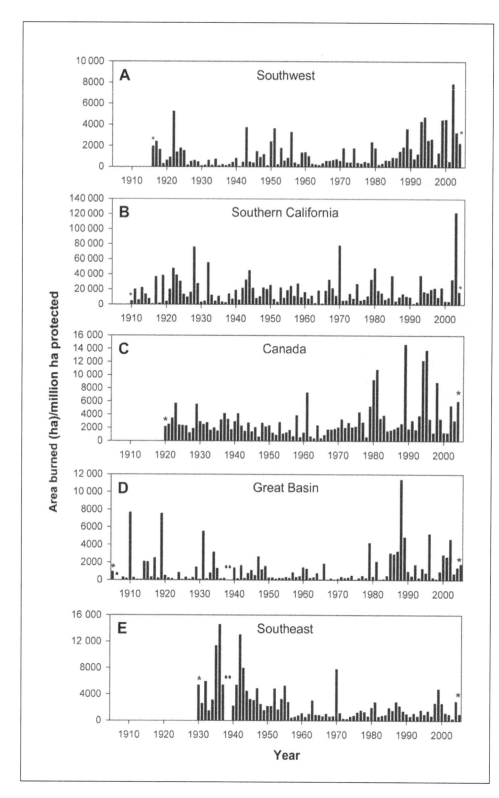

Figure 8—Historical patterns of burning. Because the area over which these data have been drawn tends to increase over time, these statistics are presented in units of hectares burned per million ha protected for (A) the Southwest, including Arizona and New Mexico private, state and federal lands (data compiled by Anthony Westerling, Scripps Institute, University of California, San Diego, from various federal databases), (B) southern California all state and federal lands for fires greater than 40 ha (data from the California Statewide Fire History database, California Department of Forestry and Fire Protection), (C) Canada, (data from Canada Large Fire Database, Canadian Forest Service), (D) Great Basin, U.S. Forest Service Intermountain Region, and (E) Southeast, U.S. Forest Service Southern Region (data from annual National Forest Fire Reports and National Interagency Fire Center). • = years of missing data. * = first and last years of available data.

areas burned during these centuries than during the 20th century (Swetnam and Baisan 1996, 2003). Similarly, the Biscuit Fire (southwestern Oregon, 2002) burned 200 000 ha, but two fires nearly twice that size occurred in the region in the mid-19th century (Walstad et al. 1990). In a similar vein, the large 2003 fires in chaparral of southern California resulted in a season with the highest area burned for the 20th century (fig. 8b), but several large fire events occurred in the 19th century (Keeley and Fotheringham 2003). For example, the Los Angeles Times (1887) reported a massive fire centered in Orange County, and Barrett (1935) provided a firsthand account of this event, which he described as the largest fire during his 33-year U.S. Forest Service career, a career that included the 93 000-ha 1932 Matilija Fire.

Assessing whether or not there have been recent changes in fire severity is difficult owing to the lack of mapped data on high-severity burns that occurred before the 20th century and lack of detailed age structure and patch size data for most forest stands (Baker and Ehle 2003). Regardless, some studies in the Southwest suggest that large crown fires were absent or rare in pure or dominant ponderosa pine forests before ca. 1900. These interpretations are based on documentary and photographic searches and comparisons (Cooper 1960), and tree age structure and fire history analyses (e.g., Barton et al. 2001, Brown and Wu 2005, Fulé et al. 2004b, Savage 1991). In some recent fires in the Southwest, e.g., the Cerro Grande, Rodeo-Chediski, and Hayman Fires, high-severity burn patches sometimes exceeded 2000 ha, which is considered outside the historical range of variability for this forest type (Allen et al. 2002, Covington and Moore 1994, Romme et al. 2003b). In contrast, there are age structure data from ponderosa mixed-conifer forests in South Dakota Black Hills, northern Colorado, and southern Idaho indicative of historical fire events dominated by crown fires (Brown et al. 1999, Ehle and Baker 2003, Kaufmann et al. 2000, Pierce et al. 2004, Sherriff 2004, Shinneman and Baker 1997). However, using age structure data to make such assessments is complicated by the evidence that even-aged ponderosa pine cohorts could be caused by episodic recruitment associated with climate changes (Brown and Wu 2005). Moreover, these studies have not clarified what the distribution of crown fire patch sizes were in the past.

Savage and Mast (2005) noted that because of the large and heavy seed of ponderosa pine, erratic seed crop production, and low success of germination and survival of seedlings, it appears that the large canopy holes (i.e., many patches 100 to 1000 ha) created by certain 20th-century fires have in some cases experienced little or no regeneration for more than 50 years. Therefore, if similar large crown fires occurred in the Southwest in the 18th or 19th centuries, they may still be visible as in-filling of canopy holes, but such events and locations have not yet been identified.

There is considerable documentary and paleoecological evidence that large, severe fires were the typical fire type in other Western ecosystems. Subalpine forests in the Rocky Mountains have historically burned in crown fires at intervals of 300 to 400 years (Buechling and Baker 2004, Despain and Romme 1991, Romme 1982). Past high-severity crown fire events can be partially reconstructed for boreal forests from stand age-structure analysis (e.g., Johnson and Gutsell 1994). Charcoal deposition studies in coastal southern California indicate that large fire events have occurred at the present frequency for at least the last 500 years (Mensing et al. 1999), although there is no evidence that these fires differed in severity from contemporary fires.

Absent old fire-scarred trees and appropriate depositional environments, it has been much more difficult to reconstruct presettlement fire regimes in the Eastern United States with any precision. Abrams (2003) and Nowacki and Abrams (2008) presented evidence for (decadal or less) frequent surface fires through much of the region now dominated by pine-oak and oak-hickory forest. Subsequent land clearing and agriculture have altered much of this landscape, and fires are almost certainly less frequent today than in the past (Delcourt and Delcourt 1998; Nowacki and Abrams, in press).

> **In Southwestern ponderosa pine there has been an increase in area burned annually and the maximum size of fires during the past century.**

To sum up, the answer to the question of whether or not fire regimes are outside the historical range of variability in recent years differs among ecosystems and regions. In Southwestern ponderosa pine there has been an increase in area burned annually and the maximum size of fires during the past century. The size of recent high-severity patches appears to be anomalous, at least on time scales of 300 to 500 years (the typical maximum ages of these forests), although this evidence has been questioned (Baker 2006b, c.f. Fúle et al. 2006). In the Great Basin, fine fuel loads from cheatgrass invasion appear to be responsible for increased fire frequency (Knapp 1996), suggesting that fire severity has possibly decreased as area burned increased (fig. 8d). Regions such as the Pacific Northwest and southern California have experienced large high-severity fires on many occasions throughout the 19th and 20th centuries so there is little evidence that the size and intensity of fires has changed (Agee 1993, Keeley et al. 1999). However, in southern California, there has been a substantial increase in fire frequency (fig. 9). The Southeast (fig. 8e) likewise exhibits little evidence of a recent increase in fire activity or fire severity.

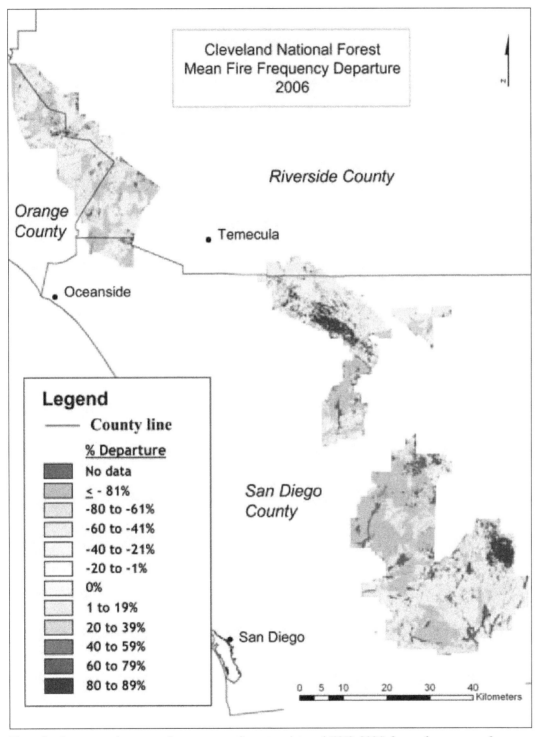

Figure 9—Percentage departure of current mean fire-return interval (1910–2006) from reference mean fire-return interval (pre-Euro-American settlement) in the Cleveland National Forest, California. Areas with negative departures (e.g., lowland chaparral and sage scrub) are experiencing more frequent fire today than in the presettlement period. Areas with positive departures (e.g., high elevation yellow pine) are experiencing less fire today than in the presettlement period. The presettlement fire-return interval is assumed to be chaparral fire-return interval assumed to be 65 years in chaparral, 75 years in coastal sage scrub is 75, and 10 years in Jeffrey pine (from Hugh Safford and Mark Borchert, U.S. Forest Service).

## Human Impacts on Fire Regimes

Land management practices—including livestock grazing, logging, fire suppression, human-caused ignitions, alien plant introductions, and habitat fragmentation owing to roads, timber harvest, and agriculture—individually and in combination have influenced fire regimes. Figure 10 illustrates how these factors interact to affect ecosystems. Fire suppression is often assumed to be of paramount importance in determining fire behavior, but on many landscapes, other factors are far more important. In some cases, timber harvest has been a bigger factor in increasing fire hazard; in other cases, grazing or alien species have been of greater importance. On some landscapes (e.g., southern California), human-caused ignitions during severe fire weather and inadequate land planning are the primary threats.

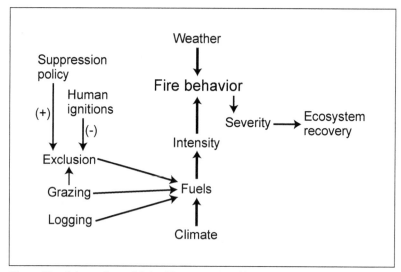

Figure 10—Schematic model that illustrates the effects of climate and fuels on fire behavior and subsequent ecosystem impacts.

In surface fire regimes, livestock grazing can greatly diminish fire frequency. Intensive livestock grazing in the Southwest (Savage and Swetnam 1990, Swetnam and Baisan 1996, Swetnam et al. 2001), parts of the Sierra Nevada (Vankat 1977), and in the Intermountain region (Heyerdahl et al. 2001, Miller and Rose 1999) has contributed to altered fire regimes since the late 19$^{th}$ century, well before effective fire suppression. Similarly, in Jeffrey pine forests of northern Baja California, fires occurred at 5- to 10-year return intervals, but declined sharply around 1790 (Stephens et al. 2003). These authors attributed this decline to the introduction of livestock grazing and cessation of burning by Native Americans (box 3), but these changes in land use are not readily parsed out from climate changes that occurred during this same period (Kitzberger et al. 2001).

## Box 3.
### Native American Influence on Fire Regimes

Colonization of North America by humans started after the last Pleistocene glacial maximum, roughly 12,000 to 14,000 years BP. There is good evidence that humans very early in their colonization altered natural ecosystems by causing or contributing to the demise of many large (>100 kg) herbivores (Martin and Klein 1984). These early Native Americans also potentially altered ecosystems by disrupting the natural fire regime through additional ignitions (Wells 1962), which potentially increased frequency of ignitions and altered seasonality of fire.

The extent to which humans disrupted the natural fire regime likely differed markedly across the continent. There is some level of agreement that they played a significant and ubiquitous role in eastern North American forested landscapes (Denevan 1992, Vale 2000). However, their effects in the West are more contentious, with some arguing for a minimal role and others for a greater role in ecosystem patterns and fire distribution (Barrett and Arno 1982, Barrett et al. 2005, Keeley 2002, Vale 2002).

This topic is relevant because some have proposed basing ecosystem management in part on a historical context that includes burning by Native Americans. Schmidt et al. (2002) and Hann et al. (2004) have included burning by Native Americans in historical reconstructions that establish baselines by which deviation of modern fire regimes from historical range of variability (HRV) (box 1) are determined. These authors contend that human subsidy of fire has affected plant evolution, and although no evidence exists to support this claim, there are indicators that burning by Native Americans has affected distribution and abundance of some plant species (Stewart 2002).

Arguments for and against including burning by Native Americans in the natural or historical fire regime (Keeley and Stephenson 2000) include:

Arguments for:
- These ignitions were part of the pre-Euro-American environment.
- Native Americans were "in tune" with their environment and managed landscapes in a responsible manner unlike contemporary humans.
- Native Americans were a "natural" part of the landscape.
- In some Western forests, burning by Native Americans was insufficient to alter burning caused by lightning, and therefore inclusion has little effect on reconstructions of historical burning patterns and the **cause** of ignition is irrelevant to the patterns and processes that sustained biodiversity historically.

Arguments against:
- Sustainable ecosystem management goals require a shift in emphasis from pre-Euro-American ideals to conditions more resilient to environmental change.
- Native Americans exploited their environment in a manner that was not qualitatively different from contemporary humans, and given sufficient time they were capable of causing unwanted changes in their environments. If management of fire is based on past Native American burning patterns, then should management of other resources (e.g., wildlife and fish) also be based on past usage by Native Americans?
- This Euro-centric perspective presumes the existence of unknown qualities that separate Native Americans from the rest of humanity.

*(continued on next page)*

> - Lightning ignitions alone are insufficient to account for fire-scar records or historical patterns of burning in some areas, and therefore inclusion is highly relevant to how we interpret fire histories.
>
> The importance of whether or not to include burning by Native Americans in the reconstruction of natural (box 1) fire regimes differs among regions and fire regimes. Fire regimes with frequent surface fires and well-developed fire histories potentially have a historical record that combines both natural fires and burning by Native Americans. Sorting out the relative contribution of each is important to the correct interpretation of these historical patterns. Crown fire ecosystems generally lack a detailed record of past fires, and thus identifying and quantifying fire source is less compelling.

**Fire intensity and fuel consumption are substantially greater when fire is returned to places where grazing has caused herbaceous fuels to be replaced by woody fuels.**

Besides diminishing fuels, livestock grazing reduces grass competition for woody species and thus enhances the recruitment of pines, which contributes to dense thickets of saplings (Arnold 1950, Belsky and Blumenthal 1997). Grazing also appears to have altered forest structure and channel erosion since the late 19$^{th}$ century (Leopold 1924), because fire intensity and fuel consumption are substantially greater when fire is returned to places where grazing has caused herbaceous fuels to be replaced by woody fuels (Zimmerman and Neuenschwander 1983). Grazing has been present much longer than fire suppression throughout western North America, and because 70 percent of Western U.S. wildlands are currently grazed (Fleischner 1994), it should be considered a widespread factor affecting fire regimes.

Past logging practices have usually not excluded fire, but in some cases have created hazardous fuel conditions commonly attributed to fire suppression (fig. 10). In some forests with mixed-severity fire regimes, fire severity may be affected more by past logging operations (owing to residual slash fuels) than fire suppression (Odion et al. 2004, Weatherspoon and Skinner 1995). For example, logging slash was a major factor in the 1871 Peshtigo Fire (Wisconsin) that burned 500 000 ha and killed over 1,200 people (Frelich 2002). Logging in and of itself is not a means of reducing fire hazard, unless slash fuels are removed or treated, either by burning or chipping (Peterson et al. 2005, Stephens 1998). However, logging can increase fire hazard owing to changes in forest composition as well as replacement of older fire-resistant trees with younger successional stages (Laudenslayer and Darr 1990, Stephens 2000) that can more readily propagate crown fire (Edminster and Olsen 1996) and increase fire severity (Agee and Huff 1987). Without surface fuel treatment, logging can increase fire intensity through its influence on insolation and surface windspeeds, leading to drier fuels and potentially more extreme fire behavior (Weatherspoon and Skinner 1995).

Timber harvest complicates our ability to make inferences about the effects of fire suppression on fire behavior. Ponderosa pine forests throughout the Western United States have been particularly targeted, and most accessible forests have been cut at least once (Ball and Schaefer 2000). For example, in northern Arizona, over 1000 km of railroads provided access for logging of large areas of forests (Stein 1993). As a result, forests that were once composed of widely spaced, old trees have been replaced by dense stands in which 98 percent of the trees are less than 100 years of age (Waltz et al. 2003). Thus, altered forest structure that contributes to fire hazard cannot be solely attributed to fire suppression. As early as the 1930s, it was evident that fires were much more common prior to fire suppression in logged areas of western Montana (Barrows 1951). The Rodeo-Chediski Fire was unusually large with a substantial level of high-severity burning, and although historical fire suppression activities played a role in altering fuel structure, logging, through its effects on fuels, insolation, and subsequent regeneration effects, may have been a factor in both the size and severity of that fire (Morrison and Harma 2002). Before this fire, much of the area had been logged one or more times, including some locations of the highest fire severity. The same can be said of the Biscuit Fire (Harma and Morrison 2003) and fires in the Klamath Mountains (Odion et al. 2004), where logged areas composed a larger portion of the high-severity burned area.

Fire spread, particularly in surface or mixed surface and crown fire regimes, is greatly disrupted by fragmentation of natural environments. Fuel disruptions owing to roads, trails, and other infrastructure may pose significant barriers to fire spread (Chang 1999). The disruption is often disproportionate to the actual size, and sometimes as little as 10 percent disruption of land cover can result in as much as 50 percent decline in fire extent (Duncan and Schmalzer 2004). This is less of a disrupting influence in crown fire ecosystems, where fires are often driven by extreme winds.

## Effects of Fire Exclusion on Forest and Shrubland Structure

Changes in ecosystem structure have the most immediate impact on fire management options, although altered fire regimes have a plethora of effects on ecosystem processes (box 4). In Southwestern ponderosa pine and oak savannas (table 1), historically frequent fire maintained a continuous understory of herbaceous fuels. This fuel distribution favored low-intensity surface fires that in turn suppressed woody plant invasion. Thus, fire maintained a distinct bimodal vertical distribution of foliage (i.e., surface and tree canopies) that resulted in a fuel gap, which limited the opportunities for surface fire to be carried into tree crowns. Fire exclusion increased surface fuels by one to two orders of magnitude and tree stem densities

*Box 4.*
## Effects of Altered Fire Regimes on Ecosystem Processes

Alteration of fire regimes has implications for sustainable ecosystem management. The consequences differ considerably among fire regimes, as well as with the history of management activities.

**The carbon cycle.** The effects of fire exclusion on forest carbon dynamics have not been studied in detail. In the short term, such exclusion leads to increased storage of carbon in accumulating fuels. However, the extensive and very intense wildfires that may eventually occur as a consequence of this fuel accumulation oxidize large quantities of carbon, and might conceivably diminish average carbon storage in the long term (van der Werf et al. 2004, Zimov et al. 1999). Either fire or mechanical harvesting reduce carbon storage. In ecosystems where fire frequency increases, carbon storage capacity is reduced.

**Nutrient cycling.** Fire exclusion can result in accumulation of nutrients in fuels, with a larger proportion of the total nutrient capital found in relatively nondecomposable coarser materials (Boerner 1982, Christensen 1977, MacKenzie et al. 2004). Burning in many ecosystem types increases the availability of soil nutrients (e.g., Christensen 1973, Sackett and Haase 1998), which may account in part for increased growth often observed in trees and understory herbs immediately following fire. Withholding fire from such systems may exacerbate nutrient limitations on plant growth. However, adding fire at too high a frequency can have deleterious long-term effects on nitrogen cycling (Carter and Foster 2004, DeLuca and Zouhar 2000, Wright and Hart 1997). These generalizations refer to more nutrient-limited ecosystems and may not be applicable to more nutrient-rich forests (Boerner et al. 2004).

**Hydrologic flows and erosion.** Increased runoff and associated erosion following fire are well documented in many ecosystems (Cannon 2001, Kirchner et al. 2001, Swanson 1981). Where fire exclusion has produced fuel accumulations and fires that are outside the historical range of variability (HRV), stream channels have suffered from patterns of erosion and sedimentation that also may be outside the HRV, although longer term perspectives place doubt on this conclusion for some landscapes (Kirchner et al. 2001). Fire suppression in some areas may be denying hydrologic events and sediment relocation that is important to long-term watershed health (e.g., Meyer 2004). On landscapes in which fire frequency is higher than it was historically (e.g., fig. 9), it has increased the long-term sediment load from watersheds (e.g., Loomis et al. 2003).

> **Community changes.** Fire exclusion can result in lower diversity and loss of rarer elements in longleaf pine (Christensen 1981, Walker and Peet 1983), ponderosa pine (Covington and Moore 1994), and mixed-conifer forests (Battles et al. 2001, Keeley et al. 2003). In addition, loss of reproduction of shade-intolerant trees occurs in deciduous (Abrams and Nowacki 1992) and coniferous forests (Cooper 1960, Harvey et al. 1980). Increased shade and increased woody litter can reduce postfire diversity patterns and, in some cases, create more uniform fuels and reduced postfire spatial variability (Rocca 2004). However, some fire-prone ecosystems are resilient to long fire-free periods that fall outside the historical range (Keeley et al. 2005b).
>
> **Landscape changes.** Landscape patch dynamics at large spatial scales can be disrupted by removal of fire (Baker 1994). This can affect animal habitat, with the greatest effects on species that depend on landscape heterogeneity to provide a suitable range of habitats for breeding, foraging, and cover (Smith 2000). Decreased landscape heterogeneity can alter fuel patterns such that fuels are distributed more homogeneously and resultant fires burn in more coarse-grained patterns, although there are notable exceptions (Turner et al. 1989).

from <125 per ha to >2500 per ha (Moore et al. 2004, Robertson and Bowser 1999, Sackett et al. 1996). Live fuels retain more water than herbaceous fuels through much of the year and are actually less flammable, meaning that drier conditions are required for their ignition. This situation facilitates the continued invasion and growth of woody plants and increased vertical continuity of fuels that can carry fire into tree crowns (fuel ladders). Thus, while fire risk may be diminished, fire hazard is increased (see box 2), and fires are potentially of higher intensity and severity (Fulé et al. 2004b).

Savannas and some grasslands may exhibit conversion to woodlands and forest with effective fire suppression. This is particularly evident on the eastern edge of the Great Plains where woodland elements historically restricted to riparian areas have expanded into adjacent grasslands (Abrams 1992, Rice and Penfound 1959). Nowacki and Abrams (2008) argued that fire suppression has facilitated successional changes in many eastern forests that have greatly diminished fire risk and fire hazard. They present evidence that presettlement oak-pine and oak-hickory forests were much more open and savanna-like than their modern counterparts. The absence of fire has facilitated the ingrowth of shade-tolerant deciduous tree species with features such as high wood and leaf lignin content and water-retaining structures (e.g., flat leaves that form a compact forest floor) that make them highly nonflammable.

Mixed-severity fire regimes include mixed-conifer forests at higher elevations in the northern Rocky Mountains, and mid-elevation forests on the Pacific slope. Historically, fire occurred every few decades, and although surface fires dominated the fire regime, the landscape comprised a mosaic of fire-induced cohorts initiated by patchy high-intensity crown fires (Fulé et al. 2003, Stephenson et al. 1991). Fire exclusion on these landscapes has resulted in less deviation from the historical range of variation in fire-return intervals than it has in surface-fire regimes, and thus less of this landscape experienced structural changes that fundamentally alter fire regime. The primary ecological change in these forests is the potential for fuel accumulation to create larger patches of crown fire (fig. 11).

Fuels in forests with mixed-severity fire regimes consist of litter, duff, and fallen branches. Accumulation of these fuels is evident after prescription burning in old-growth forests where fires have been excluded for much of the 20$^{th}$ century (fig. 12). Fire markedly reduces duff and woody fuels, and woody fuels recover within the first decade to roughly prefire levels, but duff accumulation is substantially slower (Keifer et al. 2006). Ingrowth of understory saplings and immature trees provides additional fuel as well as fuel ladders. For example, fire exclusion in

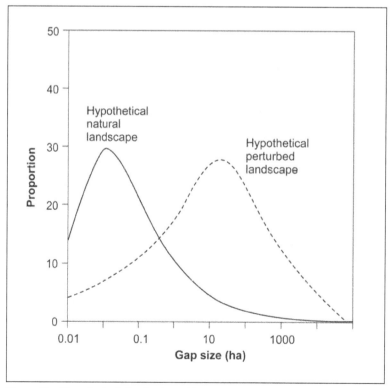

Figure 11—Hypothetical distribution of fire-generated gaps expected in forests with mixed-severity fire regimes under natural conditions, and systems perturbed by fire suppression (from Keeley and Stephenson 2000).

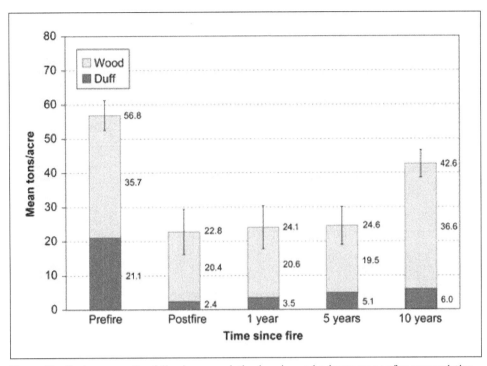

Figure 12—Fuel consumption following prescription burning and subsequent postfire accumulation in giant sequoia-mixed-conifer forests of the southern Sierra Nevada, California (mean and standard deviation bars, n = 7; from Keifer 1998). The prefire surface-fuel loads are within the range typically reported for fire regimes with return intervals of 35 to 100 years (e.g., Kauffman and Martin 1989).

Sierra Nevada giant sequoia mixed-conifer forests has increased the density of small-diameter white fir (Barbour et al. 2002, Parsons and DeBenedetti 1979) and less structural variability in tree size and distribution pattern (Taylor 2004), and the density of small-diameter trees is greatly reduced when fire is returned to these forests (fig. 13).

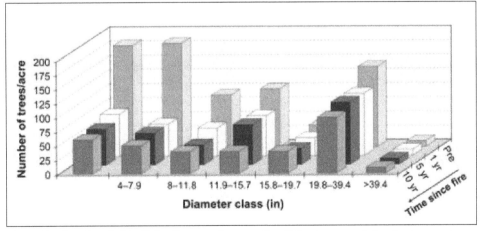

Figure 13—Density of white fir by diameter class over time following prescription fire in sequoia-mixed-conifer forests of the southern Sierra Nevada, California (Keifer 1998).

> Besides structural changes, fire exclusion results in compositional changes that differ across a moisture gradient, often favoring less fire-tolerant species.

It is often presumed that fire exclusion produces conditions in old-growth forests that make them susceptible to high-severity fires with very high mortality of overstory trees. Increased tree mortality is sometimes recorded when surface fires are successfully reintroduced in forests where fires have been excluded for long periods as a consequence of overheating of roots in deep forest floor accumulations (Fulé et al. 2004a). However, high mortality of canopy trees is not always the case, as seen after prescription fires in giant sequoia mixed-conifer forests (fig. 13) or wildfires (Odion and Hanson 2006) in the Sierra Nevada, and in Douglas-fir mixed-conifer forests in northern California (Odion et al. 2004).

Besides structural changes, fire exclusion results in compositional changes that differ across a moisture gradient, often favoring less fire-tolerant species (box 5). Ponderosa pine forests at the arid end of the gradient typically exhibit

---

### Box 5.
### Fire-Tolerance Terminology

Low-intensity surface fires are sometimes called "non-lethal" fires. This terminology appropriately describes effects on mature trees, but is of minimal value in understanding the ecological effects of fire. Surface fire regimes typically do not kill most of the larger trees, but may be lethal to seedlings, saplings, shrubs, and herbs.

"Fire tolerant," "fire sensitive," "fire dependent," and "fire adapted" are terms often applied to different tree species in mixed-conifer forests. They describe relative differences between species, but those responses differ across the landscape. For example, white fir is often termed fire intolerant or fire sensitive relative to ponderosa pine. This is most relevant in arid systems where ponderosa pine dominates in the face of frequent fire. Excluding fire from these landscapes allows the establishment of shade-tolerant white fir in the understory. When fire does occur, white fir typically experiences high mortality and is **fire sensitive** relative to pines. However, on more mesic and productive sites, white fir is the natural dominant despite the presence of frequent fires. Although seedlings can regenerate in the understory, recruitment is enhanced following fire (Mutch and Parsons 1998), and thus on these sites white fir may be considered **fire tolerant**.

**Fire dependent** refers to the necessity for postfire conditions for seedling recruitment. In this sense, white fir is clearly not fire dependent, but species such as giant sequoia are correctly termed fire dependent. Of course this term requires consideration of species within the context of communities or ecosystems. For example, in ponderosa pine savannas seedling recruitment can occur independently of fire, and dense thickets of young trees can convert these landscapes to closed-canopy forests where further recruitment is fire dependent. The related term **fire adapted** carries with it assumptions about trait origins and should be used with this understanding. The primary limitation of this term is that species in fact are not "fire adapted" as much as being adapted to particular fire regimes. For example, thick-barked oaks are often called fire adapted, but strictly speaking they are adapted to frequent surface fires, whereas thin-barked oaks may be equally fire adapted to crown fire regimes (Zedler 1995).

large changes, as higher tree density shades out further reproduction by that species but favors more shade-tolerant species such as white fir and Douglas-fir (Fulé et al. 1997). More subtle changes in composition were reported during the last half of the 20$^{th}$ century in old-growth Sierra Nevada mixed-conifer forests in more mesic locations (Ansley and Battles 1998, Roy and Vankat 1999). This is not surprising because ponderosa forests have missed more fire cycles than have mixed-conifer forests with mixed-severity fire regimes.

Exclusion of fire from forests with mixed-severity regimes has potentially increased fuel homogeneity on scales ranging from hillsides to large landscapes. Although it is often presumed that this has favored fires with more uniform fire behavior and effects, data demonstrating diminished heterogeneity are lacking. Also, heterogeneity of burning is controlled by a combination of fuel distribution, weather, and topography. Crown scorch patterns after prescription burning in California mixed-conifer forests unburned for 125 years show that such fuel conditions do not produce homogeneous fire effects (fig. 14).

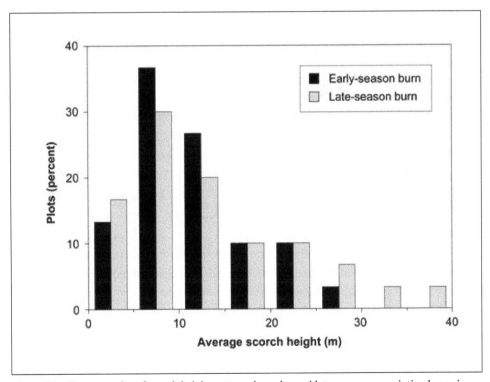

Figure 14—Heterogeneity of scorch height patterns in early- and late-season prescription burns in forests dominated by white fir, Sierra Nevada, California, following 125 years without fire (n = 30) (Knapp and Keeley 2006).

The primary disruption of fire regimes in natural crown fire ecosystems such as California chaparral shrublands has been **increased fire frequency** (fig. 9), resulting in the conversion of some portions of the landscape from native shrublands to alien herb-grasslands (box 6). In general, chaparral has not experienced the extended fire-free periods necessary for elevated fuel accumulations (Moritz 2003, Moritz et al. 2004). However, it has been suggested that the pattern of fuel distribution has become more homogeneous owing to the replacement of lightning-ignited fires, which historically would have created small patchy burns, with massive Santa Ana wind-driven fires that are most often ignited by humans (Minnich and Chou 1997). Such a change in fire size is considered unlikely to occur naturally owing to the low rate of natural lightning ignitions in this region (fig. 7). Estimates of historical burning potential suggest that without Santa Ana wind-driven fires, the rotation interval would likely have been very long, exceeding the lifespan of most shrubland species (Keeley and Fotheringham 2003). In addition, fuel mosaics, which Minnich and Chou (1997) contended are what

---

*Box 6.*
## Effects of Fuel Manipulations on Alien Invasion

Alien plant species can disrupt fire regimes either by increasing or decreasing fire activity (Brooks et al. 2004). In Western U.S. forests, effective fire suppression appears to provide some measure of resistance to alien invasion (Keeley et al. 2003), whereas forest restoration directed toward returning historical fire regimes may, under some circumstances, favor alien annual invaders (Bradley and Tueller 2004, Crawford et al. 2001, Korb et al. 2005). Historical fires occurred on a landscape that lacked the presence of alien species, many of which can spread following disturbance. In some instances the problem may require prescriptions tailored to reduce alien invasion. Grazing history, alien distribution patterns, treatment size, and fire severity are all factors that might be manipulated to reduce the alien threat linked to necessary fuel reduction projects (Keeley 2006).

Historical use of prescription fire for type conversion in crown fire shrublands such as California chaparral and Great Basin sagebrush has played a role in the widespread increase of annual grasses in these ecosystems. Fuelbreaks pose a special risk because they promote alien invasion along corridors into wildland areas (Merriam et al. 2006), and they have lower fire intensity, which promotes alien seed bank survivorship. In one comparison of ponderosa pine forests, thinning plus burning produced significantly greater alien plant abundance than burning alone (Dodson 2004).

determined this historical patchwork of burning, would have been eliminated by just a single lightning-ignited fire that lasted a week or more and carried over until a Santa Ana wind event (Zedler and Seiger 2000).

Fire exclusion has not affected fire-return intervals in Gambel oak-dominated petran chaparral of the southern Rocky Mountains and some relatively productive areas on the Colorado Plateau that develop dense piñon-juniper forests (rather than open woodlands). These systems are characterized by infrequent, severe fires occurring at intervals of many centuries (Floyd et al. 2000). Stand structure, composition, and fire behavior have apparently not been substantially altered by fire suppression (Baker and Shinneman 2004, Romme et al. 2003a). Where piñon-juniper woodland occurs at the ecotone with ponderosa pine, surface fires burning every 10 to 20 years apparently limited the piñon and juniper trees to rocky microsites (Kaye and Swetnam 1999). Recent historical changes differ, but tree densities and fuels have likely increased in some places owing to fire suppression. At the low-productivity end of the range of piñon-juniper in the Southwest, sparse, stunted woodlands occur across extensive arid landscapes, and fire appears to occur only as isolated lightning-ignited burns around individual trees or small groups (Gottfried et al. 1995).

Infrequent stand-replacing crown fires typify many cool, moist forests. These fires occur under extreme weather conditions and burn without regard to the mosaic of patch ages on the landscape (Fryer and Johnson 1988, Johnson and Fryer 1987, Turner et al. 1989). Because the fire-return interval often equals or exceeds the period of contemporary fire exclusion, it is unlikely that fire suppression has greatly altered the condition of these landscapes (Noss et al. 2006, Veblen 2003). Examples of these include subalpine forests (Buechling and Baker 2004, Masters 1990), boreal forests (Johnson et al. 1998, Weir et al. 2000), some mixed-conifer forests of the Pacific Northwest (Agee 1993, Hessburg and Agee 2003), and much of the Eastern deciduous forest (Runkle 1985). In the case of some subalpine forests in the Rocky Mountains, fire frequency has increased during the $20^{th}$ century (Sherriff et al. 2001).

## Effectiveness of Fire Suppression

The history and effectiveness of fire suppression in excluding fire differ considerably among ecosystems, landscapes, and regions. Formal policies and management protocols to suppress wildfires in the Western United States were put in place on public lands in 1911, immediately after the large fires of 1910 (Pyne 1982). Thus, one might argue that active fire suppression on public lands has been in place for nearly a century. Such management was immediately effective in areas of ready access, where fires could be discovered early and resources deployed quickly to extinguish them. In more remote areas over much of the West, suppression policies had minimal effect on fire behavior until fire towers, lookout systems, and roads in the 1930s facilitated early fire detection and deployment of firefighters. The U.S. Forest Service smoke jumping program was not used extensively until after 1945 (Cermak 2005, Pyne 1982). Thus, in more remote areas, suppression has altered fire regimes for <60 years (e.g., Whitlock et al. 2004).

The extent to which fire suppression has affected ecosystems is linked to fire regime and land use practices such as grazing and logging, as discussed above. In many western North American coniferous forests, firefighting policies have been highly effective, and many landscapes historically exposed to frequent fires have had fires suppressed for a century or more. The effect of this policy, coupled with other land management practices, is shown by fire histories in Southwestern U.S. ponderosa pine forests, wherein forests that had frequent fires until the late $19^{th}$ century show a nearly total hiatus in burning in the $20^{th}$ century (fig. 5). On these landscapes, intensive livestock grazing (usually by very large numbers of sheep) was typically the initial cause of fire regime disruptions, but active fire suppression by government agencies became a primary reason for fire exclusion after livestock numbers were greatly reduced after the 1920 (Swetnam and Baisan 2003). Disruptions of fire regimes in other parts of the Western United States followed various combinations of elimination of Native American burning practices, livestock introductions, and fire suppression efforts (e.g., Agee 1993, Arno 1980, Pyne 1982, Swetnam and Baisan 2003), whereas disruptions in Southern U.S. forests and woodlands probably related to a more complex history of human-set fires and land uses, landscape fragmentation, and fire suppression (Guyette and Spetich 2003).

It appears that fire exclusion in many conifer forests has resulted in numerous fire cycles (relative to historical frequency) being missed. However, this is not universal, and more remote forests with mixed-fire regimes did not experience fire exclusion until near the middle of the $20^{th}$ century (Whitlock et al. 2004). This is also the case for northern Mexico, where fire suppression was not practiced through much of the $20^{th}$ century (Stephens et al. 2003; Swetnam and Baisan 1996, 2003). In some mixed-conifer forests of the Pacific Northwest, fire suppression does not appear to have reduced fire activity until after the midpoint of the $20^{th}$ century (Weisberg and Swanson 2003). Inferences about the effects of fire suppression in these forests are complicated by a complex mixed-severity fire regime that involves infrequent crown fires and surface fires (Agee 1993, Hessburg and Agee 2003, Weisberg 2004).

On southern California chaparral landscapes, fire suppression policy failed to exclude fire during the 20$^{th}$ century (fig. 8b). Fires are mostly human-caused, and the current fire rotation for these crown fire regimes is 30 to 40 years (table 1); the fire-return intervals are even shorter in wildlands surrounding urban environments (fig. 9). Although fire suppression cannot be equated with fire exclusion in this region, fire suppression has still caused some effects. Throughout the 20$^{th}$ century, this fire regime has been dominated by human-caused fires that have steadily increased over time. Fire suppression has prevented large-scale conversion from native shrublands to alien grasslands, which would be expected if all human-ignited fires were allowed to burn (Keeley 2001).

Boreal forests also have a crown fire regime, but fire suppression likely has not been effective at altering the historical fire-return interval (Bridge et al. 2005, Johnson et al. 2001, Ward et al. 2001). Prefire climate sufficient to dry fuels for extended periods is a major factor determining fire activity, and because lightning is the major source of ignition in boreal forests, humans have had only local effects (Nash and Johnson 1996).

## Fire Management and Ecosystem Restoration

The objectives of restoration are typically to retain functional integrity and in some cases to maintain ecosystems within a specified range of structural and process characteristics (box 1). Fire managers intervene before fire incidence because there is a widely held belief that large fires experienced throughout western North America in recent years are the result of changes in fuel quantity and structure, and that these fires could have been prevented by better fuel management practices. These conclusions have led to initiatives such as the National Fire Plan (USDA USDI 2001), which emphasizes aggressive management of fuels as a necessary condition for sustainable resource management. These activities target a spectrum of goals that range from thinning forests and increasing wildland fire use for fire hazard reduction to more holistic ecosystem restoration. The objectives of hazard reduction are typically to alter fire behavior, reduce the severity of fire effects, and, in some cases, improve effectiveness of fire suppression. In crown fire regimes (e.g., chaparral and some boreal forests), fuel accumulation has not been the cause of large fires, and ecosystems are often within their HRV; thus there is limited need for ecosystem restoration.

**The objectives of hazard reduction are typically to alter fire behavior, reduce the severity of fire effects, and, in some cases, improve effectiveness of fire suppression.**

## Effectiveness of Prescription Burning

Prescription burning in forests with a surface-fire regime that have missed fire cycles is typically done with the objective of reducing dead and living understory fuels, for resource benefit or increased human safety, or both. This use of fire has a long history beginning with Native Americans (box 3) and is part of traditional land use practices by American settlers and rural residents (MacCleery 1996, Putz 2003). This type of forest management has been called "understory burning" or "light burning" and was frequently advocated as an appropriate way to manage pine forests in California during the early part of the $20^{th}$ century (Anonymous 1920, Cermak 2005, Olmsted 1911).

Managed prescription burns had their early origin as a means of enhancing game animal hunting in the Southeastern United States (Stoddard 1962), and today that region leads the U.S. national forests in area subjected to prescription burning (Cleaves et al. 2000). It has long been applied to limited areas of ponderosa pine in the Southwest (Biswell et al. 1973, Weaver 1968), and systematic application was initiated in mixed-conifer forests of Sequoia National Park in the late 1960s (Kilgore 1973).

Prescription burning can, in some cases, both restore historical ecosystem properties and decrease fire hazard. In the Southeastern United States there is evidence of major decreases in wildfire activity in treated forests (Davis and Cooper 1963) and reduced impacts of wildfires (Outcalt and Wade 2004). In Southwestern U.S. ponderosa pine forests, Wagle and Eakle (1979) and Finney et al. (2005) showed reduced fire severity in treated areas. Also, it has been shown that prescription burning alone is capable of meeting ecosystem restoration goals (based on conditions before Euro-American settlement) for tree density, species composition, and basal area in Southwestern U.S. ponderosa pine forests (Fulé et al. 2004a). After three decades of prescription burning in old-growth mixed-conifer forests of the Sierra Nevada, the U.S. National Park Service and U.S. Geological Survey documented that $19^{th}$-century forest structure can be reestablished without mechanical thinning (Keifer 1998, Knapp and Keeley 2006, Knapp et al. 2005). Because surface fuels accumulate rapidly in these productive forests, the longer term impact of prescription burning is the killing of smaller trees and production of higher crown levels, thus reducing ladder fuels (Kilgore and Sando 1975). Similar results with prescription burning have been reported for other old-growth mixed-conifer forests in the Western United States (Bastian 2002, Lansing 2002).

However, prescription burning is severely constrained in many cases by policy and regulations that limit the extent to which this management practice can be applied (box 7). For example, to reduce the possibility of escapes, prescription burning is normally not permitted during extreme weather conditions and when fuels are very dry. To reduce the effects of smoke on local communities, local regulations typically allow burning only during a relatively narrow window of weather conditions. Finally, prescription burns may not mimic lightning-ignited patterns in that they are often designed to produce homogeneous burning patterns that may not reflect the historical range of ignition patterns and heterogeneity of unburned and high-severity patches. Such heterogeneity may be critical to sustainability of vegetation diversity, tree recruitment (Keeley and Stephenson 2000), and wildlife habitat in some ecosystems.

Potential for prescription burning differs between surface-fire regimes and crown fire regimes. Low-intensity understory burning is rarely an option in crown fire ecosystems, and prescription crown fires for intact forests and shrublands

---

*Box 7.*
### Realities of Using Management Fire

Resource managers are faced with solving historical problems not of their making, while at the same time complying with legislative and regulatory requirements that guide planning and on-the-ground activities. For example, prescription burning is limited by air quality regulations, logistical challenges associated with complex land ownership patterns, political perspectives about the aesthetics of burning, and liability issues related to escaped fires (Yoder et al. 2004).

The problem of analyzing "fire-return interval departure," a requirement for many U.S. federal land managers, illustrates the complexity and constraints associated with managing fire. This type of analysis examines annual burning rates for a landscape of both managed and unmanaged fires relative to what would be expected if those landscapes operated under "natural" conditions. For example, in Sequoia-Kings Canyon National Parks (California), extensive fire histories provide a scientific record of historical range of varability (HRV) (box 1) in fire interval from which one can calculate the average annual proportion of landscape that burned in the past (Caprio and Graber 2000). Despite a long history of managed fire use in the park (both prescription burning and managed wildland fire), there is a large gap between what currently burns and the historical benchmark (fig. 15). Given the landscape pattern of resources at risk, air quality restrictions, and other constraints, it is unlikely this gap can be addressed through prescription burning. Approaches such as expanding the seasonal window of opportunity for burning are being considered, but the effects of burning at different times of year are not well understood (e.g., Knapp and Keeley 2006).

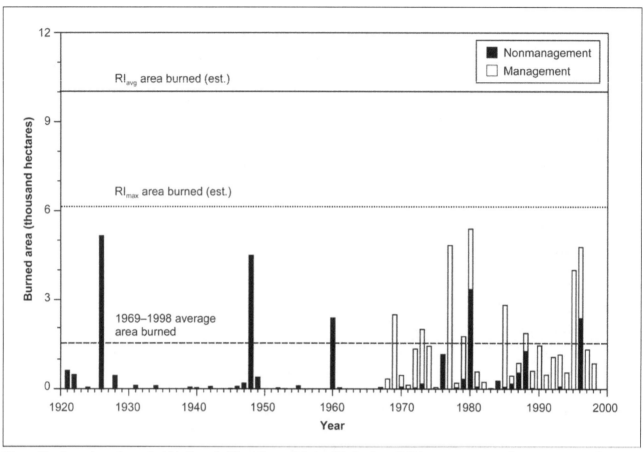

Figure 15—Annual area burned within Sequoia-Kings Canyon National Parks since 1921 by management and nonmanagement fires allowed to burn. Comparison of area burned over the last 20 years relative to estimates of area burned before Euro-American settlement is shown by horizontal lines. The largest annual area burned by management-ignited fires occurred in 1977 although the greatest number of hectares burned in any given year since 1921 was in 1980. (Caprio and Graber 2000). RI = average or maximum estimated return interval.

are challenging from an operational perspective. However, there are examples of ecosystems that have missed several fire cycles and have been managed with prescription fires. For example, Table Mountain pine in the Appalachians has serotinous cones and typically burns in high-intensity crown fires, and it has been shown that local stand-replacing prescription burning can be done safely with successful regeneration of this rare pine, although such high-intensity crown fires may not be required for successful regeneration (Waldrop et al. 2003). Sand pine, which often occurs as a sort of urban forest in Florida, also illustrates how stand-replacing prescription burning can be used successfully, in this case on fragmented landscapes (Outcalt and Greenberg 1998).

Fuel conditions in many crown fire ecosystems remain within their HRV. This applies to some Southwestern U.S. piñon-juniper woodlands where researchers have concluded that there is no ecological justification for aggressive fuel reduction

(Floyd et al. 2004). In these woodlands, the landscape is not dominated by long departures from historical fire-return intervals. Some of these ecosystems, such as some Alaskan boreal forests, can sustain prescription burning, for instance as a management tool for creating favorable wildlife habitat, without deviating from historical conditions (Vanderlinden 1996).

Despite excessively frequent fire in some places, southern California chaparral largely retains its historical composition, structure, and fire behavior, so resource benefits associated with prescription burning are limited. Nevertheless, burning is often advocated on these landscapes to decrease fire hazard. Lack of surface fuels in these shrublands means most fires are independent crown fires, and thus the goal is to maintain a landscape mosaic of young age classes with less hazardous fuels (Minnich and Chou 1997, Minnich and Dezzani 1991). Under moderate summer weather conditions, with relative humidity above 30 percent and windspeeds below 15 km per hour, fires sometimes burn out in these treated areas (Green 1981), and thus fuel treatments may limit fire spread. In any case, summer fires can be controlled before they become destructive to property. However, most large fires are ignited during the autumn foehn winds; under these severe weather conditions, fuel structure does not control fire behavior, and fires burn through, around, or over the top of these young age classes (Keeley et al. 2004). Young fuels do burn at lower fire intensity, and thus they may provide defensible space for firefighters; however, the fires grow so quickly (often exceeding 10 000 ha in the first 12 hours) that by the time firefighting resources are mobilized, most firefighters are forced into defensive positions somewhere along the periphery of the wildland-urban interface. Although fuel manipulations at the wildland-urban interface provide benefit, there is little evidence that prescription burning at large spatial scales is cost effective. Similar conclusions have been drawn about the efficacy of prescription burning in reducing fire hazard from crown fires in lodgepole pine forests of Yellowstone National Park (Wyoming) (Brown 1989, Christensen et al. 1989). Analyses of the ecological and economic effectiveness of strategic application of fuel treatments are needed for other fire regimes as well (DellaSala et al. 2004).

Restoring fire to wilderness areas presents special challenges that have been met mostly with the use of wildland fire (Kilgore and Briggs 1972). Wildland fire use (as the policy is known in the United States) allows some lightning-ignited fires to burn with suppression applied only when deemed necessary for safety or other sociopolitical reasons. Wildland fire use has been successfully applied in Sequoia-Kings Canyon National Parks (Kilgore and Taylor 1979), and in the Gila Wilderness (New Mexico) where more than 60 000 ha have burned since its natural fire program was begun in the mid-1970s. Some areas have sustained as

**Despite excessively frequent fire in some places, southern California chaparral largely retains its historical composition, structure, and fire behavior, so resource benefits associated with prescription burning are limited.**

many as four burns during that period (Boucher and Moody 1998, Rollins et al. 2001). Although crown fire has created some canopy gaps (>100 ha, upper range of HRV), the forests generally appear to have been effectively thinned with surface fire, although many dense thickets were in place before burning. In the Rincon Mountain Wilderness (Arizona), wildfires and prescription fires have maintained a relatively frequent fire regime from the late $20^{th}$ century to the present, resulting in generally open stand conditions in these ponderosa pine forests.

Wildland fire use is slowly increasing in the Western United States (Stephens and Ruth 2005). Although all major federal land management agencies have wildland fire programs, to date very little of the landscape has been allowed to burn (Parsons 2000). In most areas where this is practiced, only a small fraction of all lightning-ignited fires are allowed to burn, and commonly those under severe weather are suppressed. Thus, questions remain as to the degree to which this fire management practice restores historical patterns of ecosystem structure and function (cf. Christensen 2005).

## Effectiveness of Mechanical Fuel Manipulations

Thinning treatments are a useful means of reducing fire hazard in forests with surface and mixed-fire regimes. These treatments can differ widely in the extent to which they alter subcanopy fuels (ladder fuels), canopy base height, canopy bulk density, and canopy continuity (Agee and Skinner 2005, Peterson et al. 2005). Reduction in surface fuels decreases the potential fireline intensity and flame lengths of subcanopy fires. The distance between any remaining surface fuels and the base of the overstory tree canopies (canopy base height) is an important parameter because as this increases, so does the flame length required for canopies to ignite. Effectiveness of one treatment over another is necessarily tied to management objectives that may include reducing the severity of fire effects on forest resources, providing barriers to fire spread or defensive zones for firefighters, or restoring ecosystems to a specific condition. Much of our understanding of how mechanical fuel manipulations affect forest fire behavior is based on modeling studies that simulate fire spread (Fiedler and Keegan 2003, van Wagtendonk 1996). Results have been relatively consistent in indicating the value of combined thinning and surface fuel treatment (including but not limited to burning) for reducing subsequent fire spread rates, intensity, and severity (Johnson et al. 2007, Wallin et al. 2004).

Empirical studies in ponderosa pine and mixed-conifer forests have shown that combinations of mechanical thinning and surface fuel treatment consistently reduce wildfire severity, as measured by crown scorch and crown volume loss

> **Empirical studies in ponderosa pine and mixed-conifer forests have shown that combinations of mechanical thinning and surface fuel treatment consistently reduce wildfire severity.**

(Finney et al. 2005, Omi and Martinson 2002, Pollet and Omi 2002, Raymond and Peterson 2005). Treatments that appear to affect fire behavior the most are reductions in tree density and canopy base height (Peterson et al. 2005), although thinning is not always effective at improving the latter (Lynch et al. 2000), especially if residual stand densities are >250 stems per ha (Johnson et al. 2007). Thinning is most effective when it removes understory trees, because larger overstory trees are more resistant to heat injury (Agee and Skinner 2005). In addition, shade and competition from larger trees slows the recruitment of younger trees in the understory. Forest thinning has added benefits in reducing water stress and increasing foliar nitrogen and resin levels that enhance insect resistance (Sala et al. 2005, Wallin et al. 2004). In such treatments, it is critical that both aerial and surface fuels be treated, as slash remaining on the surface may increase fire hazard (Cram et al. 2006).

Neither modeling nor empirical studies show that fuel treatments always act as a barrier to fire spread during very extreme fire weather. This was illustrated by the 2002 Hayman Fire in which some treated forests reduced fire behavior, but spotting breached treated areas during several days of severe weather (Martinson et al. 2003). In contrast, fires may burn out in treated areas under low wind conditions and less severe drought, as illustrated by the Cone Fire (California) that burned into treated forests (Nakamura 2002). Forests with less surface fuels after treatment assist fire suppression by providing safer defensible space for firefighters, even if the treated areas do not completely stop fire spread.

Mechanical thinning, often coupled with prescription burning and other forms of surface fuel treatment, is increasingly being used to reshape forests to more closely resemble the age structure and composition of presettlement conditions based on empirically determined reference conditions (Covington and Moore 1994, Moore et al. 2004). These projects are capable of setting forests on a trajectory toward those conditions, but initial treatments typically cannot completely return forests to their original condition (Waltz et al. 2003). Mechanical thinning followed by prescription fire is an economical means of handling slash, an effective means of pruning lower branches on overstory trees, and may produce ecosystem responses similar to natural fire (Fulé et al. 2002). Physical removal of slash from thinned sites is also used to reduce surface fuels, and although it is more expensive than prescription burning, it does not affect air quality unless it is also burned offsite.

Fuelbreaks are a special class of fuel manipulation that generally comprise a broad swath of fuel reduction that runs across an otherwise untreated landscape. The effectiveness of fuelbreaks remains a matter of debate (Agee et al. 2000). They seldom represent barriers to fire spread, but zones of reduced fuels generate lower fire intensities during active wildfires, and can be used as anchor points for igniting burnout fires (to remove fuels as a fire suppression tactic) or from which prescription burning can be conducted to treat larger areas. Even for cases where there is some proven value to treated areas, the question of cost effectiveness remains (box 8).

*Box 8.*
## Economic Considerations

Cost effectiveness is critical to decisions about fire management practices (Kline 2004), with a central issue being the extent to which fuel treatments reduce suppression expenditures and subsequent wildfire dangers. For example, "minimization of cost + net value change" is a model of wildfire optimization that stresses the importance of evaluating costs in the context of economic efficiency (Donovan and Rideout 2003). We have made rapid progress in the area of relating fuel treatments to subsequent fire behavior, but gaps persist in relating these treatments to effects on forest and shrubland resources, values at risk, and human safety.

Mechanical harvest is often the preferred means of fuel reduction and forest restoration on landscapes where it is logistically feasible. Costs are a major factor in planning for and implementing fuel treatments, and prescriptions focused on reducing fire hazard may not be supported by commercial markets (Barbour et al. 2004). Removal of small trees yields relatively little volume, and the operational cost may exceed the market value (Lynch 2001, USDA FS 2005). Harvesting larger trees is one way to make these operations pay for themselves (Fiedler et al. 2004), but large gaps may promote recruitment of new saplings that require subsequent treatment. In addition, removal of larger trees is inconsistent with sustainable management for late-seral structure and for fire resistance of the residual overstory.

The costs of passive management are evident on some landscapes in the extent of large crown fires that exceed all but the rarest historical events. Fuel manipulations on these landscapes can facilitate increased resilience and sustainability to future disturbances. At the same time, fuel manipulations can cause collateral damage to soils and aquatic systems and, in some cases, promote alien plant invasion (Bisson et al. 2003, Rhodes and Odion 2004). Resource damage also occurs on other landscapes from frequent fires that degrade native ecosystems and enhance alien plant invasion. Careful analysis is required to determine the appropriate frequency, intensity, and extent of fuel manipulations for achieving specific resource objectives while minimizing negative impacts. Fire regime characteristics provide the ecological context needed to evaluate management alternatives for different landscapes.

## Ecosystem Effects of Mechanical Harvesting Versus Fire

Creation or maintenance of historical ecosystem structure and processes, or both, are typically an objective of ecosystem restoration. Mechanical harvest of trees emulates one component of natural fire by reducing the number of smaller living stems in forests (McRae et al. 2001, Perera et al. 2004). However, it does not have the same effects as fire with respect to surface fuels, understory vegetation, soils, nutrient cycling, hydrology, patch size, and snag production (Gallant et al. 2003, Kauffman 2004). In boreal forests, wildfires create more landscape heterogeneity because fire frequency is controlled by fuel moisture and, as a result, fire frequency differs by slope, aspect, and other topographic variation (McRae et al. 2001). This is difficult to emulate by harvesting trees.

Diversity and successional trajectories appear to differ for mechanically treated versus burned forests in some cases (Metlen et al. 2004) but not in others (Wienk et al. 2004). In some boreal forests, fire increases plant species diversity through duff reduction more than does tree removal (Rees and Juday 2002). Lack of duff removal by logging may result in reduced eastern white pine recruitment in Midwestern forests that have been harvested rather than burned (Weyenberg et al. 2004). In one comparison of ponderosa pine forests, thinning plus burning produced significantly higher alien plant abundance than burning alone (box 6).

## Applications in Science-Based Resource Management

This report provides an ecological foundation for management of the diverse ecosystems and fire regimes of North America. Our primary focus has been on prefire management and the range of responses required for management of diverse fire-affected ecosystems:

**Potential management options and goals need to be consistent with current and past fire regimes of specific ecosystems and landscapes.** Fire regimes differ widely among regions and among ecosystems within a region. A "one-size fits all" policy will not adequately address management goals for broad regions or multiple ecosystems within a region. Restoring and maintaining long-term sustainability and health of fire-affected systems requires management objectives and strategies that are adapted to and consistent with the fire regimes of targeted ecosystems. Options for fire management strategies may in some cases be generalized within fire regime types. For example, practical and ecologically appropriate options clearly differ among forests with surface fire regimes, forests and shrublands with crown fire regimes, and grasslands.

> Restoring and maintaining long-term sustainability and health of fire-affected systems requires management objectives and strategies that are adapted to and consistent with the fire regimes of targeted ecosystems.

**The effects of past management activities differ among ecosystems and fire regime types.** Where fire exclusion has led to fuel loads in excess of the HRV (box 1), as in some dry forests in western North America, the severity and extent of wildfires has been increasing and fuel reduction may be essential to ecological restoration. Other systems, such as California chaparral, where the balance of ignitions and suppression has led to minimal alteration of fuel loads and fire regimes, may not be good candidates for fuel treatments. In ecosystems where grazing and invasive grasses have altered fire regimes, it may be more appropriate to focus restoration efforts on reducing invasive species.

**Differences in fire history and land use history affect fuel structures and landscape patterns and can influence management options, even within a fire regime type.** Fuel structures at different spatial scales determine potential fire behavior and fire effects and are affected by succession, disturbance (including fire), and dominant use of a particular landscape (timber production, grazing, etc.). For example, differences exist between dry forest ecosystems with surface fire regimes, because surface fuels may be dominated by grasses and herbs (dry forest dominated by ponderosa pine) versus woody litter (mesic forest dominated by mixed conifer). The history of livestock grazing, as modified by interannual variations in climate, may have greater effects on surface fuels in ponderosa pine forests than in mixed-conifer forests, although the history of harvest activities may have greater effects in mixed conifer. The spatial juxtaposition of different fire histories and land use creates a mosaic of potential fire behaviors, fire effects, and habitats. None of these factors affects ecosystems with crown fire regimes nearly as much as they affect ecosystems with surface-fire regimes.

**The relative importance of fuels, climate, and weather differ among regions and ecosystems within a region; these differences greatly affect management options.** Regardless of the fire regime, large uncontrollable fires are always associated with severe fire weather. The extent to which prefire fuel manipulations can alter the course of such fires differs with the fire regime. For ecosystems such as longleaf pine or southwestern ponderosa pine, fire hazard increases when management activities that interrupt natural fire cycles lead to high fuel accumulation. For other ecosystems such as chaparral, periods of extreme fire hazard occur in most years, and severe fires are a function of human ignitions occurring under severe fire weather. Fire prevention activities and better land planning and implementation of community protection strategies may be the greatest assets to managers in these ecosystems.

**Plant species in fire-affected ecosystems may be poorly adapted to alterations in fire regimes.** Some plant species are adapted to survive and reproduce under a particular fire regime. Changes in fire frequency, severity, or seasonality that affect key ecosystem characteristics can limit the ability of those species to survive fire or to regenerate after fire. For example, when surface fire-dominated regimes are replaced by crown fire regimes in dry conifer forests, high mortality of the dominant tree species can remove the seed source needed for postfire regeneration. In chaparral vegetation, changes in fire seasonality can lead to reduced germination or seedling survival of shrubs with heat-stimulated germination. In desert shrublands and grasslands, increases in fire frequency can favor invasive annual grasses, which compete with native species and provide fuel for future fires.

**The effects of patch size must be evaluated within the context of fire regime and ecosystem characteristics.** Fire and other disturbances help to create a mosaic of vegetation with different age, structure, and fuels. Large crown fires in historical crown fire ecosystems generally do not pose a major obstacle to vegetative recovery owing to endogenous mechanisms for regeneration. In contrast, large crown fires in forests with surface-fire regimes may inhibit regeneration that depends on survival of patches of parent seed trees within dispersal distance to the fire-induced gap. The latter systems are in greatest need of management intervention before and after large fires, if the objective is to retain vegetation and structure associated with a low-severity fire regime.

**Fire severity and ecosystem responses are not necessarily correlated.** Historical fire regimes in some ecosystems are characterized by high-severity fires that kill most aboveground vegetation. Such fires may be necessary for reproduction of key species and for maintaining long-term ecosystem health, such as in chaparral and in closed-cone pine forests. In grasslands, fire severity is always high, but fire recycles nutrients and stimulates regeneration from underground plant parts. Ecosystem effects of severe fires are either neutral or positive in these situations.

**Appropriate options for forest fuel manipulations differ within the context of vegetation structure, management objectives, and economic and societal values.** Different ecosystems have different options in terms of potential fuel treatments that would reduce fire hazard. Mechanical harvest reduces ladder fuels but generally increases surface fuels unless there is further treatment. Mechanical harvest of hazardous fuels is often not cost-effective, and commercial extraction may require removal of larger trees that provide fire resistance and animal habitat. Prescription burning can consume surface fuels and increase crown base heights, often at relatively low cost, but is less efficient at removing standing fuels. Even where prescription burning may be the most cost-effective means of reducing fire hazard, it may not be feasible owing to constraints such as air quality regulations and adjacent values at risk. Strategies for reduction of hazardous fuels are more likely to be successful if short- and long-term objectives are clearly stated relative to resource values and desired conditions, and if effectiveness of all fuel treatments is monitored over time.

**Fuel manipulations alter fire behavior but are not always reliable barriers to fire spread.** The value of hazardous fuel reduction for modifying fire behavior (e.g., from crown fire to surface fire) and fire effects (e.g., tree mortality) has been documented primarily in forests with low- and mixed-severity fire regimes. Fuel treatments in these forests may diminish resource damage and provide defensible space for fire suppression activities. Their effectiveness depends on strategic location, size, and residual fuelbed structure. Most fuel treatments do not inhibit fire spread completely, especially when fuels are very dry and weather is very severe.

**Understanding historical fire patterns provides a foundation for fire management, but other factors are also important for determining desired conditions and treatments.** Management of fire regimes is more likely to be successful if it is compatible with ecosystem sustainability, feasible in the context of past disturbances and management activities, and consistent with meeting societal needs for products and values. Resource use by early North Americans influenced fire regimes in many landscapes, but was not necessarily oriented toward the ecological and resource values for which those systems are managed today. Wildland ecosystems are affected by additional and novel ecosystem stresses such as invasive species, ecosystem fragmentation, and changing climate. Desired resource conditions and fire regimes are, to a great extent, a function of management objectives such as maintaining biodiversity, increasing animal habitat, protecting the functional integrity of ecosystems, reducing alien plant invasion, maintaining water supplies, and protecting local communities. Restoration of a particular historical condition of an ecosystem as an independent objective is rarely compatible with attaining these multiple objectives. Nevertheless, knowledge of historical processes and dynamics is valuable for understanding ecosystems and identifying recent changes that are extraordinary, and which may be incompatible with species or habitat preservation.

**A variety of anthropogenic changes in climate, landscapes (e.g., fragmentation) and ecological communities (e.g., invasive species) will likely alter future fire regimes.** Flexible adaptive management that recognizes the potential for regional variation in how fire regimes respond to these global changes will be most successful. Projected climate change poses one of the more significant challenges because there is good reason to expect both direct impacts on increased fire activity as well as indirect impacts through changes in plant distribution and ecosystem fuel structure.

## Acknowledgments

We thank the Ecological Society of America, U.S. Geological Survey, The Wilderness Society, and The Nature Conservancy for financial support.

This publication was reviewed by five anonymous reviewers selected by the Ecological Society of America.

## English Equivalents

| When you know: | Multiply by: | To find: |
| --- | --- | --- |
| Kilometers (km) | 0.621 | Miles |
| Hectares (ha) | 2.47 | Acres |
| Kilowatts per meter (Kw/m) | .289 | British thermal units per foot per second |
| Pascals (pa) | .000145 | Pounds per square inch |
| Kilograms (kg) | .0011 | Tons |
| Megagrams per hectare (Mg/ha) | .446 | Tons per acre |
| Tress per hectare | .405 | Trees per acre |

## Common and Scientific Names[1]

| Common name | Scientific name |
| --- | --- |
| American chestnut | *Castanea dentata* (Marsh.) Borkh. |
| Cheatgrass | *Bromus tectorum* L. |
| Chestnut blight fungus | *Cryphonectria parasitica* (Murrill) M.E. Barr |
| Chestnut oak | *Quercus prinus* L. |
| Douglas-fir | *Pseudotsuga menziesii* (Mirb.) Franco |
| Eastern white pine | *Pinus strobus* L. |
| Gambel oak | *Quercus gambelii* (Nutt.) |
| Giant sequoia | *Sequoiadendron giganteum* (Lindl.) J. Buchholz |
| Great Basin sagebrush | *Artemisia tridentata* Nutt. |
| Longleaf pine | *Pinus palustris* Mill. |
| Mountain laurel | *Kalmia latifolia* L. |
| Pitch pine | *Pinus rigida* Mill. |
| Ponderosa pine | *Pinus ponderosa* C. Lawson |
| Red maple | *Acer rubrum* L. |
| Sand pine | *Pinus clausa* (Chapm. ex Englm) Vasey ex Sarg. |
| Table Mountain pine | *Pinus pungens* Lamb. |
| Tulip poplar | *Liriodendron tulipifera* L. |
| White fir | *Abies concolor* (Gord. & Glend.) Lindl. ex Hildebr. |

[1] Source: USDA NRCS 2008.

## Literature Cited

**Anonymous. 1920.** California divided on light burning. The Timberman. January, 1920: 36, 81.

**Abrams, M.D. 1992.** Fire and the development of oak forests. BioScience. 42: 346–353.

**Abrams, M.D. 1998.** The red maple paradox. BioScience. 48: 355–364.

**Abrams, M.D. 2003.** Where have all the white oak gone? BioScience. 53: 927–939.

**Abrams, M.D.; Nowacki, G.J. 1992.** Historical variation in fire, oak recruitment and post-logging accelerated succession in central Pennsylvania. Bulletin of the Torrey Botanical Club. 119: 19–28.

**Agee, J.K. 1993.** Fire ecology of Pacific Northwest forests. Washington, DC: Island Press.

**Agee, J.K. 1997.** The severe weather wildfire—Too hot to handle? Northwest Science. 71: 153–156.

**Agee, J.K.; Bahro, B.; Finney, M.A.; Omi, P.H.; Sapsis, D.B.; Skinner, C.N.; van Wagtendonk, J.W.; Weatherspoon, C.P. 2000.** The use of shaded fuelbreaks in landscape fire management. Forest Ecology and Management. 127: 55–66.

**Agee, J.K.; Huff, M.H. 1987.** Fuel succession in a western hemlock/Douglas-fir forest. Canadian Journal of Forest Research. 17: 697–704.

**Agee, J.K.; Skinner, C.N. 2005.** Basic principles of forest fuel reduction treatments. Forest Ecology and Management. 211: 83–96.

**Albini, F.A. 1976.** Estimating wildfire behavior and effects. Gen. Tech. Rep. INT-30. Ogden, UT: U.S. Department of Agriculture, Forest Service, Intermountain Forest and Range Experiment Station. 92 p.

**Alexander, M.E. 1982.** Calculating and interpreting forest fire intensities. Canadian Journal of Botany. 60: 349–357.

**Allen, C.D.; Breshears, D.C. 1998.** Drought-induced shift of a forest-woodland ecotone: rapid landscape response to climate variation. Proceedings of the National Academy of Sciences of the United States of America. 95(25): 14839–14842.

**Allen, C.D.; Savage, M.; Falk, D.A.; Suckling, K.F.; Swetnam, T.W.; Schulke, T.; Stacey, P.B.; Morgan, P.; Hoffman, F.; Klingel, J.T. 2002.** Ecological restoration of southwestern ponderosa pine ecosystems: a broad perspective. Ecological Applications. 12: 1418–1433.

**Andrews, P.L. 1986.** BEHAVE: fire behavior prediction and fuel modeling system: BURN subsystem, Part 1. Gen. Tech. Rep. INT-194. Ogden, UT: U.S. Department of Agriculture, Forest Service, Intermountain Forest and Range Experiment Station. 130 p.

**Ansley, J.-A.S.; Battles, J.J. 1998.** Forest composition, structure, and change in an old-growth mixed conifer forest in the northern Sierra Nevada. Bulletin of the Torrey Botanical Club. 125: 297–308.

**Arno, S.F. 1980.** Forest fire history in the northern Rockies. Journal of Forestry. 78: 460–465.

**Arnold, J.F. 1950.** Changes in ponderosa pine bunchgrass ranges in northern Arizona resulting from pine regeneration and grazing. Journal of Forestry. 48: 118–126.

**Bahre, C.J. 1985.** Wildfire in southeastern Arizona between 1859 and 1890. Desert Plants. 7: 190–194.

**Baker, W.L. 1994.** Restoration of landscape structure altered by fire suppression. Conservation Biology. 8: 763–769.

**Baker, W.L. 2006a.** Fire and restoration of sagebrush ecosystems in the Western United States. Wildlife Society Bulletin. 34: 177–185.

**Baker, W.L. 2006b.** Fire history in ponderosa pine landscapes of Grand Canyon National Park: Is it reliable enough for management and restoration? International Journal of Wildland Fire. 15: 433–437.

**Baker, W.L.; Ehle, D.S. 2001.** Uncertainty in surface-fire history: the case of ponderosa pine forests in the Western United States. Canadian Journal of Forest Research. 31: 1205–1226.

**Baker, W.L.; Ehle, D.S. 2003.** Uncertainty in fire history and restoration of ponderosa pine forests in the Western United States. In: Omi, P.N.; Joyce, L.A., eds. Fire, fuel treatments, and ecological restoration: conference proceedings. Proceedings RMRS-P-29. Fort Collins, CO: U.S. Department of Agriculture, Forest Service, Rocky Mountain Research Station: 319–333.

**Baker, W.L.; Shinneman, D.J. 2004.** Fire and restoration of piñon-juniper woodlands in the Western United States: a review. Forest Ecology and Management. 189: 1–21.

**Ball, J.J.; Schaefer, P.R. 2000.** Case no. 1: one hundred years of forest management. Journal of Forestry. 98: 4–10.

**Barbour, M.; Kelley, E.; Maloney, P.; Rizzo, D.; Royce, E.; Fites-Kaufmann, J.E. 2002.** Present and past old-growth forests of the Lake Tahoe Basin, Sierra Nevada. Journal of Vegetation Science. 13: 461–472.

**Barbour, R.J.; Fight, R.D.; Christensen, G.A.; Pinjuv, G.L.; Nagubadi, R.V. 2004.** Thinning and prescribed fire and projected trends in wood product potential, financial return, and fire hazard in Montana. Gen. Tech. Rep. PNW-GTR-606. Portland, OR: U.S. Department of Agriculture, Forest Service, Pacific Northwest Research Station. 78 p.

**Barden, L.S.; Woods, F.W. 1976.** Effects of fire on pine and pine-hardwood forests in the southern Appalachians. Forest Science. 22: 399–403.

**Barrett, L.A. 1935.** A record of forest and field fires in California from the days of the early explorers to the creation of the forest reserves. San Francisco: U.S. Department of Agriculture, Forest Service.

**Barrett, S.W.; Arno, S.F. 1982.** Indian fires as an ecological influence in the northern Rockies. Journal of Forestry. 80: 647–651.

**Barrett, S.W.; Swetnam, T.W.; Baker, W.L. 2005.** Indian fire use: deflating the legend. Fire Management Today. 65(5): 31–33.

**Barrows, J.S. 1951.** Forest fires in the northern Rocky Mountains. Station Paper 28. Missoula, MT: U.S. Department of Agriculture, Forest Service, Northern Rocky Mountains Forest and Range Experiment Station. 103 p.

**Bartlein, P.J.; Hostetler, S.W.; Shafer, S.L.; Holman, J.O.; Soloman, A.J. 2003.** The seasonal cycle of wildfire and climate in the Western United States. In: Second international wildland fire ecology and fire management congress and 5$^{th}$ symposium on fire and forest meteorology. P3.9 Orlando, FL: American Meteorological Society: [Not paged]. http://ams.confex.com/ams/pdfpapers/66935.pdf. (16 January 2008).

**Barton, A.M.; Swetnam, T.W.; Baisan, C.H. 2001.** Arizona pine (*Pinus arizonica*) stand dynamics: local and regional factors in a fire-prone madrean gallery forest of southeast Arizona, USA. Landscape Ecology. 16: 351–369.

Bastian, H. 2002. The effects of prescribed fire in mixed-conifer forests of Bryce Canyon National Park. In: Sigihara, N.G.; Morales, M.E.; Morales, T.J., eds. Proceedings of the symposium: fire in California ecosystems—integrating ecology, prevention and management. Misc. Publ. 1. Berkeley, CA: Association for Fire Ecology: 91–95.

Battles, J.J.; Shlisky, A.J.; Barrett, R.H.; Heald, R.C.; Allen-Diaz, B. 2001. The effects of forest management on plant species diversity in a Sierran conifer forest. Forest Ecology and Management. 146: 211–222.

Beckage. B.; Platt, W.J.; Slocum, M.G.; Pank, B. 2003. Influence of the El Niño Southern Oscillation on fire regimes in the Florida everglades. Ecology. 84: 3124–3130.

Belsky, A.J.; Blumenthal, D.M. 1997. Effects of livestock grazing on stand dynamics and soils in upland forests of the interior West. Conservation Biology. 11: 315–327.

Bergeron, Y.; Archambault, S. 1993. Decreasing fire frequency of forest fires in the southern boreal zone of Quebec and its relationship to global warming since the end of the "Little Ice Age." Holocene. 3: 255–259.

Bessie, W.C.; Johnson, A.E. 1995. The relative importance of fuels and weather on fire behavior in subalpine forests. Ecology. 76: 747–762.

Billings, W.D. 1990. *Bromus tectorum*, a biotic cause of ecosystem impoverishment in the Great Basin. In: Woodwell, G.M., ed. The Earth in transition: patterns and processes of biotic impoverishment. New York: Cambridge University Press: 301–322.

Bisson, P.A.; Rieman, B.E.; Luce, C.; Hessburg, P.F.; Lee, D.C.; Kershner, J.L.; Reeves, G.H.; Gresswell, R.E. 2003. Fire and aquatic ecosystems of the Western USA: current knowledge and key questions. Forest Ecology and Management. 178: 213–229.

Biswell, H.H.; Kallander, H.R.; Komarck, R.; Vogl, R.J. 1973. Ponderosa pine management. Misc. Publ. 2. Tallahassee, FL: Tall Timbers Research Station. 49 p.

Blaisdell, J.P.; Murray, R.B.; McArthur, E.D. 1982. Managing intermountain rangelands--sagebrush-grass ranges. Gen. Tech. Rep. INT-134. Ogden, UT: U.S. Department of Agriculture, Forest Service, Intermountain Forest and Range Experiment Station. 41 p.

**Boerner, R.E.J. 1982.** Fire and nutrient cycling in temperate ecosystems. BioScience. 32: 187–192.

**Boerner, R.E.J.; Brinkman, J.A.; Kennedy Sutherland, E. 2004.** Effects of fire at two frequencies on nitrogen transformations and soil chemistry in a nitrogen-enriched forest landscape. Canadian Journal of Forest Research. 34: 609–618.

**Bonnet, V.H.; Schoettle, A.W.; Shepperd, W.D. 2005.** Postfire environmental conditions influence the spatial pattern of regeneration for *Pinus ponderosa*. Canadian Journal of Forest Research. 35: 37–47.

**Borchert, M.I.; Odion, D.C. 1995.** Fire intensity and vegetation recovery in chaparral: a review. In: Keeley, J.E.; Scott, T., eds. Brushfires in California wildlands: ecology and resource management. Fairfield, WA: International Association of Wildland Fire: 91–100.

**Boucher, P.F.; Moody, R.D. 1998.** The historical role of fire and ecosystem management of fires: Gila National Forest, New Mexico. In: Pruden, T.L.; Brennan, L.A., eds. Fire in ecosystem management: shifting the paradigm from suppression to prescription. Proceedings: 20$^{th}$ Tall Timbers fire ecology conference. Tallahassee, FL: Tall Timbers Research Station: 374–379.

**Bradley, T.; Tueller, P. 2004.** Microsite recovery of vegetation in a pinyon-juniper woodland. In: Sugihara, N.G.; Morales, M.E.; Morales, T.J., eds. Proceedings of the symposium: fire management–emerging policies and new paradigms. Misc. Publ. 2. Berkeley, CA: Association for Fire Ecology: 95–106.

**Bradstock, R.A.; Auld, T.D. 1995.** Soil temperatures during experimental bushfires in relation to fire intensity: consequences for legume germination and fire management in south-eastern Australia. Journal of Applied Ecology. 32: 76–84.

**Braun, E.L. 1950.** Deciduous forests of eastern North America. Philadelphia: The Blakiston Company. 596 p.

**Bridge, S.R.J.; Miyanishi, K.; Johnson, E.A. 2005.** A critical evaluation of fire suppression effects in the boreal forest of Ontario. Forest Science. 51: 41–50.

**Brooks, M.L.; D'Antonio, C.M.; Richardson, D.M.; DiTomaso, J.M.; Grace, J.B.; Hobbs, R.J.; Keeley, J.E.; Pellant, M.; Pyke, D. 2004.** Effects of invasive alien plants on fire regimes. BioScience. 54: 677–688.

**Brotak, E.A.; Reifsnyder, W.E. 1977.** An investigation of the synoptic situations associated with major wildland fires. Journal of Applied Meteorology. 16: 867–870.

**Brown, J.K. 1989.** Could the 1988 fires in Yellowstone have been avoided through prescribed burning? Fire Management Notes. 50: 7–13.

**Brown, P.M. 2006.** Climate effects on fire regimes and tree recruitment in Black Hills ponderosa pine forests. Ecology. 87: 2500–2510.

**Brown, P.M.; Kaufmann, M.R.; Shepperd, W.D. 1999.** Long-term, landscape patterns of past fire events in a montane ponderosa pine forest of central Colorado. Landscape Ecology. 14: 513–532.

**Brown, P.M.; Wu, R. 2005.** Climate and disturbance forcing of tree recruitment in a southwestern ponderosa pine landscape. Ecology. 86: 3030–3038.

**Brown, T.J.; Hall, B.L.; Westerling, A.L. 2004.** The impact of twenty-first century climate change on wildland fire danger in the Western United States: an applications perspective. Climatic Change. 62: 365–388.

**Buechling, A.; Baker, W.L. 2004.** A fire history from tree rings in a high-elevation forest of Rocky Mountain National Park. Canadian Journal of Forest Research. 34: 1259–1273.

**Byram, G.M. 1959.** Combustion of forest fuels. In: Davis, K.P., ed. Forest fire: control and use. New York: McGraw-Hill: 61–89.

**Cannon, S.H. 2001.** Debris-flow generation from recently burned watersheds. Environmental and Engineering Geoscience. 7: 321–341.

**Caprio, A.C. 2004.** Temporal and spatial dynamics of pre-EuroAmerican fire at a watershed scale, Sequoia and Kings Canyon National Parks. In: Sugihara, N.G.; Morales, M.E.; Morales, T.J., eds. Proceedings of the symposium: fire management: emerging policies and new paradigms. Misc. Publ. 2. Berkeley, CA: Association for Fire Ecology: 91–95.

**Caprio, A.C.; Graber, D.M. 2000.** Returning fire to the mountains: Can we successfully restore the ecological role of pre-EuroAmerican fire regimes to the Sierra Nevada? In: Cole, D.N.; McCool, S.F.; O'Loughlin, J., eds. Wilderness science in a time of change conference. Proceedings RMRS-P-15. Missoula, MT: U.S. Department of Agriculture, Forest Service, Rocky Mountain Research Station: 233–241. Vol. 3.

**Carter, M.C.; Foster, C.D. 2004.** Prescribed burning and productivity in southern pine forests: a review. Forest Ecology and Management. 191: 93–109.

Cary, G.J.; Keane, R.E.; Gardner, R.H.; Lavorel, S.; Flannigan, M.D.; Davies, I.D.; Li, C.; Lenihan, J.M.; Rupp, T.S.; Mouillot, F. 2005. Comparison of the sensitivity of landscape-fire-succession models to variation in terrain, fuel pattern, climate and weather. Landscape Ecology. 21: 121–137.

Cermak, R.W. 2005. Fire in the forest: a history of forest fire control on the national forests in California, 1898-1956. Forest Research R5-FR-003. San Francisco, CA: U.S. Department of Agriculture, Forest Service, Pacific Southwest Region. 442 p.

Chang, C.-R. 1999. Understanding fire regimes. Durham, NC: Duke University. 184 p. Ph.D. dissertation.

Chapman, H.H. 1932. Is the longleaf type a climax? Ecology. 13: 328–334.

Cheney, N.P. 1990. Quantifying bushfires. Mathematical Computer Modelling. 13: 9–15.

Christensen, N.L. 1973. Fire and the nitrogen cycle in California chaparral. Science. 181: 66–68.

Christensen, N.L. 1977. Fire and soil-plant nutrient relations in a pine-wiregrass savanna on the coastal plain of North Carolina. Oecologia. 31: 27–44.

Christensen, N.L. 1981. Fire regimes in southeastern ecosystems. In: Moonely, H.A.; Bonnicksen, T.M.; Christensen, N.L.; Lotan, J.E.; Reiners, W.A., eds. Proceedings of the conference fire regimes and ecosystems properties. Gen. Tech. Rep. WO–26. Washington, DC: U.S. Department of Agriculture, Forest Service: 112–136.

Christensen, N.L. 2005. Fire in the parks: a case study for change management. The George Wright Forum. 22: 12–31.

Christensen, N.L.; Agee, J.K.; Brussard, P.F.; Hughes, J.; Knight, D.H.; Minshall, G.W.; Peek, J.M.; Pyne, S.J.; Swanson, F.J.; Thomas, J.W.; Wells, S.; Williams, S.E.; Wright, H.A. 1989. Interpreting the Yellowstone Fires of 1988. BioScience. 39: 678–685.

Clark, J.S. 1988. Effect of climate change on fire regimes in northwestern Minnesota. Nature. 334: 233–235.

Clark, J.S.; Robinson, J. 1993. Paleoecology of fire. In: Crutzen, P.J.; Goldammer, J.G., eds. Fire in the environment. New York: Wiley: 193–214.

**Cleaves, D.A.; Haines, T.K.; Martinez, J. 2000.** Influences on prescribed burning activity in the National Forest System. In: Moser, W.K.; Moser, C.F., eds. Fire and forest ecology: innovative silviculture and vegetation management. Proceedings: 21$^{st}$ Tall Timbers fire ecology conference. Tallahassee, FL: Tall Timbers Research Station: 170–177.

**Collins, B.M.; Omi, P.N.; Chapman, P.L. 2006.** Regional relationships between climate and wildfire–burned area in the interior West, USA. Canadian Journal of Forest Research. 36: 699–709.

**Conard, S.G.; Weise, D.R. 1998.** Management of fire regime, fuels, and fire effects in southern California chaparral: lessons from the past and thoughts for the future. In: Pruden, T.L.; Brennan, L.A., eds. Fire in ecosystem management: shifting the paradigm from suppression to prescription. Proceedings: 20$^{th}$ Tall Timbers fire ecology conference. Tallahassee, FL: Tall Timbers Research Station: 342–350.

**Cooper, C.F. 1960.** Changes in vegetation, structure, and growth of southwestern pine forest since white settlement. Ecological Monographs. 30: 129–164.

**Covington, W.W.; Moore, M.M. 1994.** Southwestern ponderosa pine forest structure and resource conditions: changes since Euro–American settlement. Journal of Forestry. 92: 39–47.

**Cram, D.; Baker, T.; Boren, J. 2006.** Wildland fire effects in silviculturally treated vs. untreated stands of New Mexico and Arizona. Res. Pap. RMRS-RP-55. Fort Collins, CO: U.S. Department of Agriculture, Forest Service, Rocky Mountain Research Station. 28 p.

**Crawford, J.A.; Wahren, C.-H.A.; Kyle, S.; Moir, W.H. 2001.** Responses of exotic plant species to fires in *Pinus ponderosa* forests in northern Arizona. Journal of Vegetation Science. 12: 261–268.

**Crow, T.T. 1988.** Reproductive mode and mechanisms for self-replacement of northern red oak (*Quercus rubra*)–a review. Forest Science. 34: 19–40.

**Davis, L.S.; Cooper, R.W. 1963.** How prescribed burning affects wildfire occurrence. Journal of Forestry. 61: 915–917.

**Davis, M.B. 1983.** Quaternary history of deciduous forests of eastern North America and Europe. Annals of the Missouri Botanical Garden. 70: 550–563.

**DeBano, L.F. 2000.** Water repellency in soils: a historical overview. Journal of Hydrology. 231-232: 4–32.

**Delcourt, P.A.; Delcourt, H.R. 1998.** The influence of prehistoric human-set fires on oak chestnut forests in the southern Appalachians. Castanea. 63: 337–345.

**DellaSala, D.A.; Williams, J.E.; Williams, C.D.; Franklin, J.F. 2004.** Beyond smoke and mirrors: a synthesis of fire policy and science. Conservation Biology. 18: 976–986.

**DeLuca, T.H.; Zouhar, K.L. 2000.** Effects of selection harvest and prescribed fire on the soil nitrogen status of ponderosa pine forests. Forest Ecology and Management. 138: 263–271.

**Denevan, W.M. 1992.** The pristine myth: the landscape of the Americas in 1492. Annals of the Association of American Geographers. 82: 369–385.

**Despain, D.G.; Romme, W.H. 1991.** Ecology and management of high-intensity fires in Yellowstone National Park. Proceedings Tall Timbers Ecology Conference. 17: 43–57.

**DeVivo, M.S. 1991.** Indian use of fire and land clearance in the southern Appalachians. In: Nodvin, S.C.; Waldrop, T.A., eds. Fire and the environment: ecological and cultural perspectives. Gen. Tech. Rep. SE-69. Asheville, NC: U.S. Department of Agriculture, Forest Service, Southeastern Forest Experiment Station: 306–310.

**Dickinson, M.B.; Johnson, E.A. 2001.** Fire effects on trees. In: Johnson, E.A.; Miyanishi, K., eds. Forest fires: behavior and ecological effects. San Diego, CA: Academic Press: 477–525.

**Dieterich, J.H. 1980.** The composite fire interval: a tool for more accurate interpretation of fire history. In: Stokes, M.A.; Dieterich, J.H., eds. Proceedings of the fire history workshop. Gen. Tech. Rep. RM-GTR-81. Fort Collins, CO: U.S. Department of Agriculture, Forest Service, Rocky Mountain Forest and Range Experiment Station: 8–14.

**Dieterich, J.H.; Swetnam, T.W. 1984.** Dendrochronology of a fire scarred ponderosa pine. Forest Science. 30: 238–247.

**Dodge, J.M. 1972.** Forest fuel accumulation—a growing problem. Science. 177: 139–142.

**Dodson, E.K. 2004.** Monitoring change in exotic plant abundance after fuel reduction/restoration treatments in ponderosa pine forests of western Montana. Missoula: University of Montana. 95 p. M.S. thesis.

**Doerr, S.H.; Shakesby, R.A.; Blake, W.H.; Chafer, C.J.; Humphreys, G.S.; Wallbrink, P.J. 2006.** Effects of differing wildfire severities on soil wettability and implications for hydrological response. Journal of Hydrology. 319: 295–311.

**Donnegan, J.A.; Veblen, T.T.; Sibold, J.S. 2001.** Climatic and human influences on fire history in Pike National Forest, central Colorado. Canadian Journal of Forest Research. 31: 1526–1539.

**Donovan, G.H.; Rideout, D.B. 2003.** A reformulation of the cost plus net value change ($C+NVC$) model of wildfire economics. Forest Science. 49: 318–323.

**Duncan, B.W.; Schmalzer, P.A. 2004.** Anthropogenic influences on potential fire spread in a pyrogenic ecosystem of Florida, USA. Landscape Ecology. 19: 153–165.

**Early, L.S. 2004.** Looking for longleaf: the fall and rise of an American forest. Chapel Hill, NC: University of North Carolina Press. 336 p.

**Edminster, C.B.; Olsen, W.K. 1996.** Thinning as a tool for restoring and maintaining stand structure in stands of southwestern ponderosa pine. In: Covington, W.W.; Wagner, P.K., eds. Conference on adaptive ecosystem restoration and management: restoration of cordilleran conifer landscapes of North America. Gen. Tech. Rep. RM-GTR-278. Fort Collins, CO: U.S. Department of Agriculture, Forest Service, Rocky Mountain Forest and Range Experiment Station: 61–67.

**Ehle, D.S.; Baker, W.L. 2003.** Disturbance and stand dynamics in ponderosa pine forests in Rocky Mountain National Park, USA. Ecological Monographs. 73: 543–566.

**Elliott, K.J.; Hendrick, R.L.; Major, A.E.; Vose, J.M.; Swank, W.T. 1999.** Vegetation dynamics after prescribed fire in the southern Appalachians. Forest Ecology and Management. 114: 199–213.

**Falk, D.A.; Swetnam, T.W. 2003.** Scaling rules and probability models for surface fire regimes in ponderosa pine forests. In: Omi, P.N.; Joyce, L.A. Fire, fuel treatments, and ecological restoration: conference proceedings. Proceedings RMRS-P-29. Fort Collins, CO: U.S. Department of Agriculture, Forest Service, Rocky Mountain Research Station: 301–318.

**Fiedler, C.E.; Keegan, C.E. 2003.** Reducing crown fire hazard in fire-adapted forests of New Mexico. In: Omi, P.N.; Joyce, L.A., tech. eds. Fire, fuel treatments, and ecological restoration: conference proceedings. Proceedings RMRS-P-29. Fort Collins, CO: U.S. Department of Agriculture, Forest Service, Rocky Mountain Research Station: 39–48.

**Fiedler, C.E.; Keegan, C.E., III; Woodall, C.W.; Morgan, T.A. 2004.** A strategic assessment of crown fire hazard in Montana: potential effectiveness and costs of hazard reduction treatments. Gen. Tech. Rep. PNW-GTR-622. Portland, OR: U.S. Department of Agriculture, Forest Service, Pacific Northwest Research Station. 48 p.

**Finney, M.A. 1995.** The missing tail and other considerations for the use of fire history models. International Journal of Wildland Fire. 5: 197–202.

**Finney, M.A. 1998.** FARSITE: Fire Area Simulator-model development and evaluation. Res. Pap. RMRS-RP-4. Ogden, UT: U.S. Department of Agriculture, Forest Service, Rocky Mountain Research Station. 47 p.

**Finney, M.A.; McHugh, C.W.; Grenfell, I.C. 2005.** Stand- and landscape-level effects of prescribed burning on two Arizona wildfires. Canadian Journal of Forest Research. 35: 1714–1722.

**Flannigan, M.D.; Logan, K.A.; Amiro, B.D.; Skinner, W.; Stocks, B. 2005.** Future area burned in Canada. Climatic Change. 72(1-2): 1–16.

**Flannigan, M.D.; Stocks, B.J.; Wotten, B.M. 2000.** Climate change and forest fires. Science of the Total Environment. 262: 221–229.

**Fleischner, T.L. 1994.** Ecological costs of livestock grazing in western North America. Conservation Biology. 8: 629–644.

**Floyd, M.L.; Hanna, D.D.; Romme, W.H. 2004.** Historical and recent fire regimes in piñon-juniper woodlands on Mesa Verde, Colorado, USA. Forest Ecology and Management. 198: 269–289.

**Floyd, M.L.; Romme, W.H.; Hanna, D.D. 2000.** Fire history and vegetation pattern in Mesa Verde National Park, Colorado, USA. Ecological Applications. 10: 1666–1680.

**Frelich, L.E. 2002.** Forest dynamics and disturbance regimes. Studies from temperate evergreen-deciduous forests. Cambridge: Cambridge University Press. 266 p.

Fryer, G.I.; Johnson, E.A. 1988. Reconstructing fire behavior and effects in a subalpine forest. Journal of Applied Ecology. 25: 1063–1072.

Fulé, P.Z.; Cocke, A.E.; Heinlein, T.A.; Covington, W.W. 2004a. Effects of an intense prescribed forest fire: Is it ecological restoration? Ecological Restoration. 12: 220–230.

Fulé, P.Z.; Covington, W.W.; Moore, M.M. 1997. Determining reference conditions for ecosystem management of southwestern ponderosa pine forests. Ecological Applications. 7: 895–908.

Fulé, P.Z.; Covington, W.W.; Smith, H.B.; Springer, J.D.; Heinlein, T.A.; Huisinga, K.D.; Moore, M.M. 2002. Comparing ecological restoration alternatives: Grand Canyon, Arizona. Forest Ecology and Management. 170: 19–41.

Fulé, P.Z.; Crouse, J.E.; Cocke, A.E.; Moore, M.M.; Covington, W.W. 2004b. Changes in canopy fuels and potential fire behavior 1880-2040: Grand Canyon, Arizona. Ecological Modelling. 175: 231–248.

Fulé, P.Z.; Crouse, J.E.; Heinlein, T.A.; Moore, M.M.; Covington, W.W.; Verkamp, G. 2003. Mixed-severity fire regime in a high-elevation forest of Grand Canyon, Arizona, USA. Landscape Ecology. 18: 465–486.

Fúle, P.Z.; Heinlein, T.A.; Covington, W.W. 2006. Fire histories in ponderosa pine forests of Grand Canyon are well supported: reply to Baker. International Journal of Wildland Fire. 15: 439–445.

Gallant, A.L.; Hansen, A.J.; Councilman, J.S.; Monte, D.K.; Betz, D.W. 2003. Vegetation dynamics under fire exclusion and logging in a Rocky Mountain watershed, 1856-1996. Ecological Applications. 13: 385–403.

Garren, K.H. 1943. Effects of fire on vegetation of the Southeastern United States. Botanical Review. 9: 617–654.

Gedalof, Z.; Peterson, D.L.; Mantua, N.J. 2005. Atmospheric, climatic and ecological controls on extreme wildfire years in the Northwestern United States. Ecological Applications. 15: 154–174.

Gill, A.M. 1973. Effects of fire on Australia's natural vegetation. In: Annual report. Canberra, ACT, Australia: Commonwealth Scientific and Industrial Research Organisation, Division of Plant Industry: 41–46.

**Gillett, N.P.; Weaver, A.J.; Zwiers, F.W.; Flannigan, M.D. 2004.** Detecting the effect of climate change on Canadian forest fires. Geophysical Research Letters. 31(18): Art. No. L18211.

**Girardin, M.P. 2007.** Interannual to decadal changes in area burned in Canada from 1781-1982 and the relationship to Northern Hemisphere land temperatures. Global Ecology and Biogeography. 16: 557–566.

**Gottfried, G.J.; Swetnam, T.W.; Allen, C.D.; Betancourt, J.L.; Chung-McCoubrey, A.L. 1995.** Pinyon-juniper woodlands. In: Finch, D.M.; Tainter, J.A., eds. Ecology, diversity, and sustainability of the Middle Rio Grande Basin. Gen. Tech. Rep. RM-GTR-268. Fort Collins, CO: U.S. Department of Agriculture, Forest Service, Rocky Mountain Forest and Range Experiment Station: 95–132. Chapter 6.

**Graham, R.T., ed. 2003.** Hayman Fire case study: summary. Gen. Tech. Rep. RMRS-GTR-115. Fort Collins, CO: U.S. Department of Agriculture, Forest Service, Rocky Mountain Research Station. 32 p.

**Green, L.R. 1981.** Burning by prescription in chaparral. Gen. Tech. Rep. PSW-51. Albany, CA: U.S. Department of Agriculture, Forest Service, Pacific Southwest Forest and Range Experiment Station. 36 p.

**Greene, D.F.; Johnson, E.A. 2000.** Tree recruitment from burn edges. Canadian Journal of Forest Research. 30: 1264–1274.

**Grissino-Mayer, H.D.; Swetnam, T.W. 2000.** Century-scale climate forcing of fire regimes in the American Southwest. Holocene. 10: 213–220.

**Guyette, R.P.; Spetich, M.A. 2003.** Fire history of oak-pine forests in the Lower Boston Mountains, Arkansas, USA. Forest Ecology and Management. 180: 463–474.

**Halsey, R.W. 2004.** Fire, chaparral, and survival in southern California. San Diego, CA: Sunbelt Publications. 188 p.

**Hann, W.; Shlisky, A.; Havlina, D.; Schon, K.; Barrett, S.; DeMeo, T.; Pohl, K.; Menakis, J.; Hamilton, D.; Jones, J.; Levesque, M.; Frame, C. 2004.** Interagency fire regime condition class guidebook. Version 1.3.0. http://frames.nbii.gov/frcc/documents/FRCC_Guidebook_08.01.17.pdf. (11 June 2008).

**Harma, K.; Morrison, P. 2003.** Analysis of vegetation mortality and prior landscape condition, 2002 Biscuit Fire Complex. 23 p. http://www.pacificbio.org/Projects/Fires/reports/VegetationMortalityAnalysis-screenres.pdf. (17 January 2008).

**Harvey, T.; Shellhammer, H.S.; Stecker, R.E. 1980.** Giant sequoia ecology: fire and reproduction. Scientific Monograph Series No. 20. Washington, DC: U.S. Department of the Interior, National Park Service. 182 p.

**Heinselman, M.L. 1981.** Fire intensity and frequency as factors in the distribution and structure of northern ecosystems. In: Mooney, H.A.; Bonnicksen, T.M.; Christensen, N.L.; Lotan, J.E.; Reiners, W.A., eds. Proceedings of the conference fire regimes and ecosystems properties. Gen. Tech. Rep. WO-26. Washington, DC: U.S. Department of Agriculture, Forest Service: 7–57.

**Hely, C.; Flannigan, M.; Bergeron, Y.; McRae, D.M. 2001.** Role of vegetation and weather on fire behavior in the Canadian mixed wood boreal forest using two fire behaviour prediction systems. Canadian Journal of Forest Research. 31: 430–441.

**Hessburg, P.F.; Agee, J.K. 2003.** An environmental narrative of inland Northwest United States forests, 1800–2000. Forest Ecology and Management. 178: 23–59.

**Hessl, A.E.; McKenzie, D.; Schellhaas, R. 2004.** Drought and Pacific Decadal Oscillation linked to fire occurrence in the inland Pacific Northwest. Ecological Applications. 14: 425–442.

**Heyerdahl, E.K.; Brubaker, L.B.; Agee, J.K. 2001.** Spatial controls of historical fire regimes: a multiscale example from the interior West, USA. Ecology. 82: 660–678.

**Heyerdahl, E.K.; Brubaker, L.B.; Agee, J.K. 2002** Annual and decadal climate forcing of historical fire regimes in the interior Pacific Northwest. Holocene. 12(5): 597–604.

**Hughes, M.K.; Diaz, H.F. 1994.** Was there a Medieval Warm Period, and if so where and when? Climate Change. 26: 109–142.

**Hurrell, J.W. 1996.** Influence of variations in extratropical wintertime teleconnections on Northern Hemisphere temperature. Geophysical Research Letters. 23: 665–668.

**Johnson, E.A. 1992.** Fire and vegetation dynamics: studies from the North American boreal forest. Cambridge: Cambridge University Press. 129 p.

**Johnson, E.A.; Fryer, G.I. 1987.** Historical vegetation change in the Kananaskis Valley, Canadian Rockies. Canadian Journal of Botany. 65: 853–858.

**Johnson, E.A.; Gutsell, S.L. 1994.** Fire frequency models, methods and interpretations. Advances in Ecological Research. 25: 239–287.

**Johnson, E.A.; Miyanishi, K. 2001.** Strengthening fire ecology's roots. In: Johnson, E.A.; Miyanishi, K., eds. Forest fires: behavior and ecological effects. San Diego, CA: Academic Press: 1–9.

**Johnson, E.A.; Miyanishi, K.; Bridge, S.R.J. 2001.** Wildfire regime in the boreal forest and the idea of suppression and fuel buildup. Conservation Biology. 15: 1554–1557.

**Johnson, E.A.; Miyanishi, K.; Weir, J.M.H. 1998.** Wildfires in the western Canadian boreal forest: landscape patterns and ecosystem management. Journal of Vegetation Science. 9: 603–610.

**Johnson, E.A.; Morin, H.; Gagnon, M.K.; Greene, D.F. 2003.** A process approach to understanding disturbance and forest dynamics for sustainable forestry. In: Burton, P.J.; Messier, C.; Smith, W.; Adomowicz, W.L., eds. Towards sustainable forest management of the boreal forest. Ottawa, Canada: NRC Research Press: 261–306.

**Johnson, E.A.; Van Wagner, C.E. 1985.** The theory and use of two fire history models. Canadian Journal of Forest Research. 15: 214–220.

**Johnson, E.A.; Wowchuk, D.R. 1993.** Wildfires in the southern Canadian Rocky Mountains and their relationship to mid-tropospheric anomalies. Canadian Journal of Forest Research. 23: 1213–1222.

**Johnson, M.C.; Peterson, D.L.; Raymond, C.L. 2007.** Guide to fuel treatments in dry forests of the Western United States: assessing forest structure and fire hazard. Gen. Tech. Rep. PNW-GTR-686. Portland, OR: U.S. Department of Agriculture, Forest Service, Pacific Northwest Research Station. 322 p.

**Kauffman, J.B. 2004.** Death rides the forest: perceptions of fire, land use, and ecological restoration of western forests. Conservation Biology. 18: 878–882.

**Kauffman, J.B.; Martin, R.E. 1989.** Fire behavior, fuel consumption, and forest-floor changes following prescribed understory fires in Sierra Nevada mixed conifer forests. Canadian Journal of Forest Research. 19: 445–462.

**Kauffmann, M.R.; Regan, C.M.; Brown, P.M. 2000.** Heterogeneity in ponderosa pine/Douglas-fir forests: age and size structure in unlogged and logged landscapes of central Colorado. Canadian Journal of Forest Research. 30: 698–711.

**Kaye, M.W.; Swetnam, T.W. 1999.** An assessment of fire, climate, and Apache history in the Sacramento Mountains, New Mexico. Physical Geography. 20: 305–330.

**Keeley, J.E. 2001.** Fire and invasive species in Mediterranean-climate ecosystems of California. In: Galley, K.E.M.; Wilson, T.P., eds. Proceedings of the invasive species workshop: the role of fire in the control and spread of invasive species. Misc. Publ. 11. Tallahassee, FL: Tall Timbers Research Station: 81–94.

**Keeley, J.E. 2002.** Native American impacts on fire regimes of the California coastal ranges. Journal of Biogeography. 29: 303–320.

**Keeley, J.E. 2004.** Impact of antecedent climate on fire regimes in coastal California. International Journal of Wildland Fire. 13: 173–182.

**Keeley, J.E. 2006.** Fire management impacts on invasive plant species in the Western United States. Conservation Biology. 20: 375–384.

**Keeley, J.E.; Fotheringham, C.J. 2003.** Impact of past, present, and future fire regimes on North American mediterranean shrublands. In: Veblen, T.T.; Baker, W.L.; Montenegro, G.; Swetnam, T.W., eds. Fire and climatic change in temperate ecosystems of the western Americas. New York: Springer: 218–262.

**Keeley, J.E.; Fotheringham, C.J.; Keeley, M.B. 2005a.** Determinants of postfire recovery and succession in mediterranean-climate shrublands of California. Ecological Applications. 15: 1515–1534.

**Keeley, J.E.; Fotheringham, C.J.; Morais, M. 1999.** Reexamining fire suppression impacts on brushland fire regimes. Science. 284: 1829–1832.

**Keeley, J.E.; Fotheringham, C.J.; Moritz, M.A. 2004.** Lessons from the October 2003 wildfires in southern California. Journal of Forestry. 102: 26–31.

**Keeley, J.E.; Lubin, D.; Fotheringham, C.J. 2003.** Fire and grazing impacts on plant diversity and alien plant invasions in the southern Sierra Nevada. Ecological Applications. 13: 1355–1374.

**Keeley, J.E.; Pfaff, A.H.; Safford, H.D. 2005b.** Fire suppression impacts on postfire recovery of Sierra Nevada chaparral shrublands. International Journal of Wildland Fire. 14: 255–265.

**Keeley, J.E.; Stephenson, N.L. 2000.** Restoring natural fire regimes in the Sierra Nevada in an era of global change. In: Cole, D.N.; McCool, S.F.; O'Loughlin, J., eds. Wilderness science in a time of change conference. Proceedings RMRS-P-15. Missoula, MT: U.S. Department of Agriculture, Forest Service, Rocky Mountain Research Station: 255–265. Vol. 3.

**Keeley, J.E.; Zedler, P.H. 1998.** Evolution of life histories in *Pinus*. In: Richardson, D.M., ed. Ecology and biogeography of *Pinus*. Cambridge: Cambridge University Press: 219–250.

**Keifer, M. 1998.** Fuel load and tree density changes following prescribed fire in the giant sequoia-mixed conifer forest: the first 14 years of fire effects monitoring. In: Pruden, T.L.; Brennan, L.A., eds. Fire in ecosystem management: shifting the paradigm from suppression to prescription. Proceedings: 20th Tall Timbers fire ecology conference. Tallahassee, FL: Tall Timbers Research Station: 306–309.

**Keifer, M.; van Wagtendonk, J.W.; Buhler, M. 2006.** Long-term surface fuel accumulation in burned and unburned mixed-conifer forests of the central and southern Sierra Nevada, CA (USA). Fire Ecology. 2: 53–72.

**Kilgore, B.M. 1973.** Impact of prescribed burning on a sequoia-mixed conifer forest. In: Proceedings, 12th Tall Timbers fire ecology conference. Tallahassee, FL: Tall Timbers Research Station: 345–375.

**Kilgore, B.M.; Briggs, G.S. 1972.** Restoring fire to high elevation forests in California. Journal of Forestry. 70: 267–271.

**Kilgore, B.M.; Sando, R.W. 1975.** Crown fire potential in a sequoia forest after prescribed burning. Forest Science. 21: 83–87.

**Kilgore, B.M.; Taylor, D. 1979.** Fire history of a sequoia-mixed conifer forest. Ecology. 60: 129–142.

**Kilgore, B.M. 1987.** The role of fire in wilderness: a state-of knowledge review. In: Lucas, R.C., comp. Proceedings—National wilderness research conference: issues, state-of-knowledge, future directions. Gen. Tech. Rep. INT-220. Ogden, UT: U.S. Department of Agriculture, Forest Service, Intermountain Research Station: 70–103.

**Kirchner, J.W.; Finkel, R.C.; Riebe, C.S.; Granger, D.E.; Clayton, J.L.; King, J.G.; Megahan, W.F. 2001.** Mountain erosion over 10 yr, 10 k.y., and 10 m.y. time scales. Geology. 29: 591–594.

**Kitzberger, T.; Brown, P.M.; Heyerdahl, E.K.; Swetnam, T.W.; Veblen, T.T. 2007.** Contingent Pacific-Atlantic ocean influence on multi-century wildfire synchrony over western North America. Proceedings of the National Academy of Sciences. 104: 543–548.

**Kitzberger, T.; Swetnam, T.W.; Veblen, T.T. 2001.** Inter-hemispheric synchrony of forest fires and the El Nino-Southern Oscillation. Global Ecology and Biogeography. 10: 315–326.

**Kline, J.D. 2004.** Issues in evaluating the costs and benefits of fuel treatments to reduce wildfire in the nation's forests. Res. Note PNW-RN-542. Portland, OR: U.S. Department of Agriculture, Forest Service, Pacific Northwest Research Station. 46 p.

**Knapp, E.E.; Brennan, T.J.; Ballenger, E.A.; Keeley, J.E. 2005.** Fuel reduction and coarse woody debris dynamics with early season and late season prescribed fires in a Sierra Nevada mixed conifer forest. Forest Ecology and Management. 208: 383–397.

**Knapp, E.E.; Keeley, J.E. 2006.** Heterogeneity in burn severity and burn pattern with early season and late season prescribed fire in a mixed conifer forest. International Journal of Wildland Fire. 15: 1–9.

**Knapp, P.A. 1996.** Cheatgrass (*Bromus tectorum* L.) dominance in the Great Basin Desert. History, persistence, and influences of human activities. Global Environmental Change. 6: 37–52.

**Knapp, P.A. 1998.** Spatio-temporal patterns of large grassland fires in the intermountain West, U.S.A. Global Ecology and Biogeography. 7: 259–272.

**Korb, J.E.; Springer, J.D.; Powers, S.R.; Moore, M.M. 2005.** Soil seed banks in *Pinus ponderosa* forests in Arizona: clues to site history and restoration potential. Applied Vegetation Science. 8: 103–112.

**Landres, P.B.; Morgan, P.; Swanson, F.J. 1999.** Overview of the use of natural variability concepts in managing ecological systems. Ecological Applications. 9: 1179–1188.

**Lansing, C. 2002.** Fire effects monitoring results in Yosemite National Park white fir-mixed conifer forest: fuel load and tree density changes. In: Sugihara, N.G.; Morales, M.E.; Morales, T.J., eds. Proceedings of the symposium: fire in California ecosystems: integrating ecology, prevention and management. Misc. Publ. 1. Berkeley, CA: Association for Fire Ecology: 364–371.

**Laudenslayer, W.F.J.; Darr, H.H. 1990.** Historical effects of logging on the forests of the Cascade and Sierra Nevada Ranges of California. Transactions of the Western Section of the Wildlife Society. 26: 12–23.

**Lenihan, J.M.; Drapek, R.; Bachelet, D.; Neilson, R.P. 2003.** Climate change effects on vegetation distribution, carbon, and fire in California. Ecological Applications. 13: 1667–1681.

**Leopold, A. 1924.** Grass, brush, timber, and fire in southern Arizona. Journal of Forestry. 22: 1–10.

**Littell, J.S. 2006.** Climate impacts to forest ecosystem processes: Douglas-fir growth in Northwestern U.S. mountain landscapes and area burned by wildfire in Western U.S. eco-provinces. Seattle: University of Washington. 171 p. Ph.D. dissertation.

**Loomis, J.; Wohlgemuth, P.; González-Cabán, A.; English, D. 2003.** Economic benefits of reducing fire-related sediment in southwestern fire-prone ecosystems. Water Resources Research. 39(9): 1260.

**Lorimer, C.G. 1980.** Age structure and disturbance history of a southern Appalachian virgin forest. Ecology. 61: 1169–1184.

**Lorimer, C.G. 1985.** Methodological considerations in the analysis of forest disturbance history. Canadian Journal of Forest Research. 15: 200–213.

**Los Angles Times. 1887.** A delayed report of forest fires. September 27.

**Lynch, D.L. 2001.** Financial results of ponderosa pine forest restoration in Southwestern Colorado. In: Vance, R.K.; Edminster, C.B.; Covington, W.W.; Blank, J.A., eds. Ponderosa pine ecosystems restoration and conservation: steps toward stewardship. Proceedings RMRS-P-22. Ogden, UT: U.S. Department of Agriculture, Forest Service, Rocky Mountain Research Station: 141–148.

**Lynch, D.L.; Romme, W.H.; Floyd, M.L. 2000.** Forest restoration in southwestern ponderosa pine. Journal of Forestry. 98: 17–24.

**MacCleery, D.W. 1996.** American forests: a history of resiliency and recovery. Forest History Society Issues Series FS-540. Durham, NC: U.S. Department of Agriculture, Forest Service. In cooperation with: The Forest History Society. 58 p.

**Mack, R.N. 1981.** Invasion of *Bromus tectorum* L. into western North America: an ecological chronicle. Agro-Ecosystems. 7: 145–165.

**MacKenzie, M.D.; DeLuca, T.H.; Sala, A. 2004.** Forest structure and organic horizon analysis along a fire chronosequence in the low elevation forests of western Montana. Forest Ecology and Management. 203: 331–343.

**Mackintosh, W.A.; Joerg, W.L.G., eds. 1935.** Canadian frontiers of settlement. Toronto: Macmillan Company of Canada Limited. 9 vol.

**Mann, M.; Bradley, R.S.; Hughes, M.K. 1998.** Global scale temperature patterns and climate forcing over the past six centuries. Nature. 392: 779–788.

**Martin, P.S.; Klein, R.G., eds. 1984.** Quaternary extinctions: a prehistoric revolution. Tucson, AZ: University of Arizona Press. 892 p.

Martinson, E.; Omi, P.N.; Shepperd, W.P. 2003. Effects of fuel treatments on fire severity. In: Graham, R.T., ed. Hayman Fire case study. Gen. Tech. Rep. RMRS-GTR-114. Fort Collins, CO: U.S. Department of Agriculture, Forest Service, Rocky Mountain Research Station: 96–126.

Masters, A.M. 1990. Changes in forest fire frequency in Kootenay National Park, Canadian Rockies. Canadian Journal of Botany. 68: 1763–1767.

McKenzie, D.; Gedalof, Z.; Peterson, D.L.; Mote, P. 2004. Climatic change, wildfire, and conservation. Conservation Biology. 18: 890–902.

McRae, D.J.; Duchesne, L.C.; Freedman, B.; Lynham, T.J.; Woodley, S. 2001. Comparisons between wildfire and forest harvesting and their implications in forest management. Environmental Reviews. 9: 223–260.

Meehl, G.A.; Washington, W.M.; Wigley, T.M.L.; Arblaster, J.M.; Dai, A.A. 2003. Solar and greenhouse gas forcing and climatic response in the 20$^{th}$ century. Journal of Climate. 16: 426–444.

Mensing, S.A.; Michaelsen, J.; Byrne, R. 1999. A 560-year record of Santa Ana fires reconstructed from charcoal deposited in the Santa Barbara Basin, California. Quaternary Research. 51: 295–305.

Merriam, K.E.; Keeley, J.E.; Beyers, J.L. 2006. Fuel breaks affect nonnative species abundance in Californian plant communities. Ecological Applications. 16: 515–527.

Metlen, K.L.; Fiedler, C.F.; Youngblood, A. 2004. Understory response to fuel reduction treatments in the Blue Mountains of northeastern Oregon. Northwest Science. 78: 175–185.

Meyer, G.A. 2004. Yellowstone fires and the physical landscape. In: Wallace, L.L., ed. After the fires: the ecology of change in Yellowstone National Park. New Haven, CT: Yale University Press: 29–51.

Millar, C.I. 1997. Comments on historical variation and desired condition as tools for terrestrial landscape analysis. In: Sommarstrom, S., ed. Proceedings of the 6$^{th}$ biennial watershed management conference. Water Resources Center Report 92. Davis, CA: University of California: 105–131.

Millar, C.I.; Wolfenden, W.B. 1999. The role of climate change in interpreting historical variability. Ecological Applications. 9: 1207–1216.

**Miller, R.F.; Eddleman, L.L. 2001.** Spatial and temporal changes of sagebrush grouse habitat in the sagebrush biome. Tech. Bull. 151. Corvallis, OR: Oregon State University, Agricultural Experiment Station. 35 p.

**Miller, R.F.; Rose, J.A. 1999.** Fire history and western juniper encroachment in sagebrush steppe. Journal of Range Management. 52: 550–559.

**Millspaugh, S.H.; Whitlock, C.; Bartlein, P.J. 2004.** Postglacial fire, vegetation, and climate history of the Yellowstone-Lamar and central plateau provinces, Yellowstone National Park. In: Wallace, L.L., ed. After the fires: the ecology of change in Yellowstone National Park. New Haven, CT: Yale University Press: 10–28.

**Minnich, R.A. 1983.** Fire mosaics in southern California and northern Baja California. Science. 219: 1287–1294.

**Minnich, R.A.; Chou, Y.H. 1997.** Wildland fire patch dynamics in the chaparral of southern California and northern Baja California. International Journal of Wildland Fire. 7: 221.

**Minnich, R.A.; Dezzani, R.J. 1991.** Suppression, fire behavior, and fire magnitudes in Californian chaparral at the urban/wildland interface. In: DeVries, J.J., ed. California watersheds at the urban interface, proceedings of the $3^{rd}$ biennial watershed conference. Davis, CA: University of California: 67–83.

**Miyanishi, K. 2001.** Duff consumption. In: Johnson, E.A.; Kiyanishi, K., eds. Forest fires: behavior and ecological effects. San Diego, CA: Academic Press: 437–475.

**Moody, J.A.; Martin, D.A. 2001.** Initial hydrologic and geomorphic response following a wildfire in the Colorado Front Range. Earth Surface Processes and Landforms. 26: 1049–1070.

**Moore, M.M.; Huffman, D.W.; Fulé, P.Z.; Covington, W.W.; Crouse, J.E. 2004.** Comparison of historical and contemporary forest structure and composition on permanent plots in southwestern ponderosa pine forests. Forest Science. 50: 162–175.

**Moreno, J.M.; Oechel, W.C. 1994.** Fire intensity as a determinant factor of postfire plant recovery in southern California chaparral. In: Moreno, J.M.; Oechel, W.C., eds. The role of fire in mediterranean-type ecosystems. New York: Springer-Verlag: 26–45.

**Morgan, P.; Hardy, C.C.; Swetnam, T.W.; Rollins, M.G.; Long, D.G. 2001.** Mapping fire regimes across time and space: understanding coarse and fine-scale fire patterns. International Journal of Wildland Fire. 10: 329–342.

**Moritz, M.A. 2003.** Spatiotemporal analysis of controls on shrubland fire regimes: age dependency and fire hazard. Ecology. 84: 351–361.

**Moritz, M.A.; Keeley, J.E.; Johnson, E.A.; Schaffner, A.A 2004.** Testing a basic assumption of shrubland fire management: Does the hazard of burning increase with the age of fuels? Frontiers in Ecology and the Environment. 2: 67–72.

**Morrison, P.; Harma, K. 2002.** Analysis of land ownership and prior land management activities within the Rodeo Chediski Fires, Arizona. http://www.pacificbio.org/Projects/Fires/reports/RodeoChediskiFires8July2002.pdf. (20 January 2008).

**Murphy, P.J.; Mudd, J.P.; Stocks, B.J.; Kasischke, E.S.; Barry, D.; Alexander, M.E.; French, N.H.F. 2000.** Historical fire records in the North American boreal forest. In: Kasischke, E.S.; Stocks, B.J., eds. Fire, climate change, and carbon cycling in the boreal forest. New York: Springer: 274–289.

**Mutch, L.S.; Parsons, D.J. 1998.** Mixed conifer forest mortality and establishment before and after prescribed fire in Sequoia National Park, California. Forest Science. 44: 341–355.

**Myers, R.L. 1985.** Fire and the dynamic relationship between Florida sandhill and sand pine scrub vegetation. Bulletin of the Torrey Botanical Club. 112: 241–252.

**Nakamura, G. 2002.** Cone Fire tests fuel reduction treatment effectiveness. Davis, CA: University of California, Cooperative Extension Forestry Program. [Not paged.]

**Nash, C.H.; Johnson, E.A. 1996.** Synoptic climatology of lightning-caused forest fires in subalpine and boreal forests. Canadian Journal of Forest Research. 26: 1859–1874.

**National Wildfire Coordinating Group. 2006.** Glossary of wildland fire terminology. Boise, ID: National Wildfire Coordinating Group, National Interagency Fire Center. 183 p.

**Norman, S.; Taylor, A.H. 2002.** Variation in fire-return intervals across a mixed-conifer forest landscape. In: Sugihara, N.G.; Morales, M.E.; Morales, T.J., eds. Proceedings of the symposium: fire in California ecosystems: integrating ecology, prevention and management. Misc. Publ. 1. Berkeley, CA: Association for Fire Ecology: 170–179.

**Noss, R.F.; Franklin, J.F.; Baker, W.L.; Schoennagel, T.; Moyle, P.B. 2006.** Managing fire-prone forests in the Western United States. Frontiers in Ecology and the Environment. 4: 481–487.

**Nowacki, G.J.; Abrams, M.D. [2008].** The demise of fire and "mesophication" of the Eastern United States. BioScience. 58: 123–138.

**Odion, D.C.; Frost, E.J.; Strittholt, J.R.; Jiang, H.; DellaSala, D.A.; Moritz, M.A 2004.** Patterns of fire severity and forest conditions in the western Klamath Mountains, California. Conservation Biology. 18: 927–936.

**Odion, D.C.; Hanson, C.T. 2006.** Fire severity in conifer forests of Sierra Nevada, California. Ecosystems. 9: 1177–1189.

**Olmsted, F.E. 1911.** Fire and the forest–the theory of "light burning." Sierra Club Bulletin. 8: 43–49.

**Omi, P.N.; Martinson, E.J. 2002.** Effect of fuels treatment on wildfire severity. http://welcome.warnercnr.colostate.edu/frws/research/westfire/FinalReport.pdf. (11 June 2008).

**Ottmar, R.D.; Sandberg, D.V.; Riccardi, C.L.; Prichard, S.J. 2007.** An overview of the Fuel Characteristic Classification System–quantifying, classifying, and creating fuelbeds for resource planning. Canadian Journal of Forest Research. 37(12): 2383–2393.

**Outcalt, K.W.; Greenberg, C.H. 1998.** A stand-replacement prescribed burn in sand pine scrub. In: Pruden, T.L.; Brennan, L.A., eds. Fire in ecosystem management: shifting the paradigm from suppression to prescription. Proceedings: 20th Tall Timbers fire ecology conference. Tallahassee, FL: Tall Timbers Research Station: 141–145.

**Outcalt, K.W.; Wade, D.D. 2004.** Fuels management reduces tree mortality from wildfires in Southeastern United States. Southern Journal of Applied Forestry. 28: 28–34.

**Parsons, D.J. 2000.** The challenge of restoring natural fire to wilderness. In: Cole, D.N.; McCool, S.F.; O'Loughlin, J., eds. Wilderness science in a time of change conference. Missoula, MT: U.S. Department of Agriculture, Forest Service, Rocky Mountain Research Station: 276–282.

**Parsons, D.J.; DeBenedetti, S.H. 1979.** Impact of fire suppression on a mixed-conifer forest. Forest Ecology and Management. 2: 21–33.

**Perera, A.H.; Buse, L.J.; Weber, M.G., eds. 2004.** Emulating natural forest landscape disturbances: concepts and applications. New York: Columbia University Press. 252 p.

**Peterson, D.L.; Johnson, M.C.; Agee, J.K.; Jain, T.B.; McKenzie, D.; Reinhardt, E.R. 2005.** Forest structure and fire hazard in dry forests of the Western United States. Gen. Tech. Rep. PNW-GTR-628. Portland, OR: U.S. Department of Agriculture, Forest Service, Pacific Northwest Research Station. 30 p.

**Pierce, J.L.; Meyer, G.A.; Jull, J.T. 2004.** Fire-induced erosion and millennial-scale climate change in northern ponderosa pine forests. Nature. 432: 87–90.

**Platt, W.J.; Evans, G.W.; Rathbun, S.L. 1988.** The population dynamics of a long-lived conifer (*Pinus palustris*). American Naturalist. 131: 491–525.

**Podur, J.; Martell, D.L.; Knight, K. 2002.** Statistical quality control analysis of forest fire activity in Canada. Canadian Journal of Forest Research. 32: 195–205.

**Pollet, J.; Omi, P.N. 2002.** Effect of thinning and prescribed burning on crown fire severity in ponderosa pine forests. International Journal of Wildland Fire. 11: 1–10.

**Putz, F.E. 2003.** Are rednecks the unsung heroes of ecosystem management? Wild Earth. 13(2/3): 10–14. http://www.afrc.uamont.edu/whited/Putz%202003.pdf. (20 January 2008).

**Pyne, S.J. 1982.** Fire in America: a cultural history of wildland and rural fire. Princeton, NJ: Princeton University Press. 654 p.

**Pyne, S.J. 2001.** Year of the fires: the story of the great fires of 1910. New York: Viking. 322 p.

**Raymond, C.L.; Peterson, D.L. 2005.** Fuel treatments alter the effects of wildfire in a mixed-evergreen forest, Oregon, USA. Canadian Journal of Forest Research. 35: 2981–2995.

**Reed, W.J.; Johnson, E.A. 2004.** Statistical methods for estimating historical fire frequency from multiple fire-scar data. Canadian Journal of Forest Research. 34: 2306–2313.

**Reed, W.J.; McKelvey, K.S. 2002.** Power-law behaviour and parametric models for the size distribution of forest fires. Ecological Modeling. 150: 239–254.

**Rees, D.C.; Juday, G.P. 2002.** Plant species diversity on logged versus burned sites in central Alaska. Forest Ecology and Management. 155: 291–302.

**Rhodes, J.J.; Odion, D.C. 2004.** Evaluation of the efficacy of forest manipulations still needed. BioScience. 54: 980.

**Rice, E.L.; Penfound, W.T. 1959.** The upland forests of Oklahoma. Ecology. 40: 593–608.

**Robertson, P.A.; Bowser, Y.H. 1999.** Coarse woody debris in mature *Pinus ponderosa* stands in Colorado. Journal of the Torrey Botanical Society. 126: 255–267.

**Rocca, M.E. 2004.** Spatial considerations in fire management: the importance of heterogeneity for maintaining diversity in a mixed-conifer forest. Durham, NC: Duke University. 142 p. Ph.D. dissertation.

**Rollins, M.G.; Swetnam, T.W.; Morgan, P. 2001.** Evaluating a century of fire patterns in two Rocky Mountain wilderness areas using digital fire atlases. Canadian Journal of Forest Research. 31: 2107-2123.

**Romme, W.H. 1982.** Fire and landscape diversity in subalpine forests of Yellowstone National Park. Ecological Monographs. 52: 199–221.

**Romme, W.H.; Floyd-Hanna, L.; Hanna, D.D. 2003a.** Ancient piñon-juniper forests of Mesa Verde and the West: a cautionary note for forest restoration programs. In: Omi, P.N.; Joyce, L.A., eds. Fire, fuel treatments, and ecological restoration: conference proceedings. Proceedings RMRS-P-29. Fort Collins, CO: U.S. Department of Agriculture, Forest Service, Rocky Mountain Research Station: 335–350.

**Romme, W.H.; Veblen, T.T.; Kaufmann, M.R.; Sherriff, R.; Regan, C.M. 2003b.** Historical (Pre-1860) and current (1860-2002) fire regimes. In: Graham, R.T., ed. Hayman Fire case study. Gen. Tech. Rep. RMRS-GTR-114. Fort Collins, CO: U.S. Department of Agriculture, Forest Service, Rocky Mountain Research Station: 181–195.

**Rothermel, R.C. 1983.** How to predict the spread and intensity of wildfires. Gen. Tech. Rep. INT-143. Ogden, UT: U.S. Department of Agriculture, Forest Service, Intermountain Forest and Range Experiment Station. 161 p.

**Rothermel, R.C. 1991.** Predicting behavior and size of crown fires in the northern Rocky Mountains. Res. Pap. INT-438. Ogden, UT: U.S. Department of Agriculture, Forest Service, Intermountain Research Station. 46 p.

**Roy, D.G.; Vankat, J.L. 1999.** Reversal of human-induced vegetation changes in Sequoia National Park, California. Canadian Journal of Forest Research. 29: 399–412.

**Runkle, J.R. 1985.** Disturbance regimes in temperate forests. In: Pickett, S.T.A.; White, P.S., eds. The ecology of natural disturbance and patch dynamics. Orlando, FL: Academic Press: 17–33.

**Sackett, S.S.; Haase, S.M. 1998.** Two case histories for using prescribed fire to restore ponderosa pine ecosystems in northern Arizona. In: Pruden, T.L.; Brennan, L.A., eds. Fire in ecosystem management: shifting the paradigm from suppression to prescription. Proceedings: 20th Tall Timbers fire ecology conference. Tallahassee, FL: Tall Timbers Research Station: 308–389.

**Sackett, S.S.; Haase, S.M.; Harrington, M.G. 1996.** Lessons learned from fire use restoring southwestern ponderosa pine ecosystems. In: Covington, W.W.; Wagner, P.K, eds. Conference on adaptive ecosystem restoration and management: restoration of cordilleran conifer landscapes of North America. Gen. Tech. Rep. RM-GTR-278. Fort Collins, CO: U.S. Department of Agriculture, Forest Service, Rocky Mountain Forest and Range Experiment Station: 54–61.

**Sala, A.; Peters, G.D.; McIntyre, L.R.; Harrington, M.G. 2005.** Physiological responses of ponderosa pine in western Montana to thinning, prescribed fire and burning season. Tree Physiology. 25: 339–348.

**Sandberg, D.V.; Ottmar, R.D.; Cushon, G.H. 2001.** Characterizing fuels in the 21st century. International Journal of Wildland Fire. 10: 381–387.

**Savage, M. 1991.** Structure dynamics of a southwestern ponderosa pine forest under chronic human influence. Annals of the Association of American Geographers. 81: 271–289.

**Savage, M.; Mast, J.N. 2005.** How resilient are southwestern ponderosa pine forests after crown fires? Canadian Journal of Forest Research. 35: 967–977.

**Savage, M.; Swetnam, T.W. 1990.** Early 19th-century fire decline following sheep pasturing in a Navajo ponderosa pine forest. Ecology. 71: 2374–2378.

**Schmidt, K.M.; Menakis, J.P.; Hardy, C.C.; Hann, W.J.; Bunnell, D.L. 2002.** Development of coarse-scale spatial data for wildland fire and fuel management. Gen. Tech. Rep. RMRS-GTR-87. Fort Collins, CO: U.S. Department of Agriculture, Forest Service, Rocky Mountain Research Station. 41 p. plus CD.

**Schoennagel, T.; Veblen, T.T.; Romme, W. 2004.** The interaction of fire, fuels, and climate across Rocky Mountain landscapes. BioScience. 54: 661–676.

**Schoennagel, T.; Veblen, T.T.; Romme, W.H.; Sibold, J.S.; Cook, E.R. 2005.** ENSO and PDO variability affect drought-induced fire occurrence in Rocky Mountain subalpine forests. Ecological Applications. 15(6): 2000–2014.

**Schroeder, M.; Glovinsky, M.; Hendricks, V.; Hood, F.; Hull, M.; Jacobson, H.; Kirkpatrick, R.; Krueger, D.; Mallory, L.; Oertel, A.; Reese, R.; Sergius, L.; Syverson, C. 1964.** Synoptic weather types associated with critical fire weather. Berkeley, CA: U.S. Department of Agriculture, Forest Service, Pacific Southwest Forest and Range Experiment Station. 492 p.

**Scott, J.H.; Reinhardt, E.D. 2001.** Assessing crown fire potential by linking models of surface and crown fire behavior. Res. Pap. RMRS-RP-29. Fort Collins, CO: U.S. Department of Agriculture, Forest Service, Rocky Mountain Research Station. 59 p.

**Sherriff, R.L. 2004.** The historic range of variability of ponderosa pine in the northern Colorado Front Range: past fire types and fire effects. Boulder, CO: University of Colorado. 220 p. Ph.D. dissertation.

**Sherriff, R.L.; Veblen, T.T.; Sibold, J.S. 2001.** Fire history in high elevation subalpine forests in the Colorado Front Range. Ecoscience. 8: 369–380.

**Shinneman, D.J.; Baker, W.L. 1997.** Nonequilibrium dynamics between catastrophic disturbances and old-growth forests in ponderosa pine landscapes of the Black Hills. Conservation Biology. 11: 1276–1288.

**Sibold, J.S.; Veblen, T.T. 2006.** Relationships of subalpine forest fires in the Colorado Front Range with interannual and multidecadal-scale climatic variation. Journal of Biogeography. 33: 833–842.

**Simard, A.J.; Haines, D.A.; Main, W.A. 1985.** Relations between El Nino/Southern Oscillation anomalies and wildland fire activity in the United States. Agricultural and Forest Meteorology. 36: 93–104.

**Skinner, W.R.; Stocks, B.J.; Martell, D.L.; Bonsal, B.; Shabbar, A. 1999.** The association between circulation anomalies in the mid-troposphere and area burned by wildland fire in Canada. Theoretical and Applied Climatology. 63: 89–105.

**Smith, J.K.E. 2000.** Wildland fire in ecosystems: effects of fire on fauna. Gen. Tech. Rep. RMRS-GTR-42. Ogden, UT: U.S. Department of Agriculture, Forest Service, Rocky Mountain Research Station. 83 p. Vol 1.

**Snyder, J.R. 1986.** The impact of wet season and dry season prescribed fires on Miami Rock Ridge pineland, Everglades National Park. South Florida Research Center Rep. No. SFRC-86/06. Homestead, FL: U.S. Department of the Interior, National Park Service, South Florida Research Center, Everglades National Park. 106 p.

**Stein, P.H. 1993.** Railroad logging on the Coconino and Kaibab National Forests, 1887 to 1966: supplemental report to a National Register of Historic Places multiple property documentation form. SWCA Archaeology Report 93-16. Report for contract no. 42-8167-2-0373. Flagstaff, AZ: SWCA, Inc.

**Stephens, S.L. 1998.** Evaluation of the effects of silvicultural and fuels treatments on potential fire behaviour in Sierra Nevada mixed-conifer forests. Forest Ecology and Management. 105: 21–35.

**Stephens, S.L. 2000.** Mixed conifer and red fir forest structure and uses in 1899 from the central and northern Sierra Nevada, California. Madroño. 47: 43–52.

**Stephens, S.L.; Ruth, L.W. 2005.** Federal forest-fire policy in the United States. Ecological Applications. 15: 532–542.

**Stephens, S.L.; Skinner, C.N.; Gill, S.J. 2003.** Dendrochonology-based fire history of Jeffrey pine-mixed conifer forests in the Sierra San Pedro Martir, Mexico. Canadian Journal of Forest Research. 33: 1090–1101.

**Stephenson, N.L. 1999.** Reference conditions for giant sequoia forest restoration: structure, process, and precision. Ecological Applications. 9: 1253–1265.

**Stephenson, N.L.; Parsons, D.J.; Swetnam, T.W. 1991.** Restoring natural fire to the Sequoia-mixed conifer forest: Should intense fire play a role? In: Hermann, S.M., ed. High-intensity fire in wildlands: management challenges and options. Proceedings: 17$^{th}$ Tall Timbers fire ecology conference. Tallahassee, FL: Tall Timbers Research Station: 17: 321–337.

**Stewart, O.C. 2002.** Forgotten fires: Native Americans and the transient wilderness. Norman, OK: University of Oklahoma Press. 352 p.

**Stocks, B.J.; Mason, J.A.; Todd, J.B.; Bosch, E.M.; Wotton, B.M.; Amiro, B.D.; Flannigan, M.D.; Hirsh, K.G.; Logan, K.A.; Martell, D.L.; Skinner, W.R. 2003.** Large forest fires in Canada, 1959–1997. Journal of Geophysical Research. 108(D1, 8149): 1–12.

**Stoddard, H.L.S. 1962.** Use of fire in pine forests and game lands of the deep southeast. In: Proceedings: 1st Tall Timbers fire ecology conference: 1: 31–42.

**Strauss, D.; Dednar, L.; Mees, R. 1989.** Do one percent of forest fires cause ninety-nine percent of the damage? Forest Science. 35: 319–328.

**Swanson, F.J. 1981.** Fire and geomorphic processes. In: Mooney, H.A.; Bonnicksen, T.M.; Christensen, N.L.; Lotan, J.E.; Reiners, W.A., eds. Gen. Tech. Rep. WO-26. Proceedings of the conference fire regimes and ecosystem properties. Washington, DC: U.S. Department of Agriculture, Forest Service: 401–420.

**Swetnam, T.W. 1993.** Fire history and climate change in giant sequoia groves. Science. 262: 885-889.

**Swetnam, T.W.; Allen, C.D.; Betancourt, J.L. 1999.** Applied historical ecology: using the past to manage for the future. Ecological Applications. 9: 1189–1206.

**Swetnam, T.W.; Baisan, C.H. 1996.** Historical fire regime patterns in the Southwestern United States since AD 1700. In: Allen, C.D., tech. ed. Fire effects in Southwestern forests: proceedings of the Second La Mesa fire symposium. Gen. Tech. Rep. RM-GTR-286. Fort Collins, CO: U.S. Department of Agriculture, Forest Service, Rocky Mountain Forest and Range Experiment Station: 11–32.

**Swetnam, T.W.; Baisan, C.H. 2003.** Tree-ring reconstructions of fire and climate history in the Sierra Nevada and Southwestern United States. In: Veblen, T.T.; Baker, W.L.; Montenegro, G.; Swetnam, T.W. Fire and climatic change in temperate ecosystems of the western Americas. New York: Springer: 158–195.

**Swetnam, T.W.; Baisan, C.H.; Kaib, J.M. 2001.** Forest fire histories in the sky islands of La Frontera. In: Webster, G.L.; Bahre, C.J., eds. Changing plant life of La Frontera: observations on vegetation in the United States/Mexico borderlands. Albuquerque, NM: University of New Mexico Press: 95–119. Chapter 7.

**Swetnam, T.W.; Betancourt, J.L. 1990.** Fire-Southern Oscillation relations in the Southwestern United States. Science. 249: 1017–1020.

**Taylor, A.H. 2004.** Identifying forest reference conditions on early cut-over lands, Lake Tahoe Basin, USA. Ecological Applications. 14: 1903–1920.

**Tchir, T.L.; Johnson, E.A.; Miyanishi, K. 2004.** A model of fragmentation in the Canadian boreal forest. Canadian Journal of Forest Research. 34: 2248–2262.

**Trenberth, K.E.; Hurrell, J.W. 1994.** Decadal atmosphere-ocean variations in the Pacific. Climate Dynamics. 9: 303–319.

**Turner, M.G.; Gardner, R.H.; Dale, V.H.; O'Neill, R.V. 1989.** Predicting the spread of disturbance across heterogeneous landscapes. Oikos. 55: 121–129.

**Umbanhowar, C.E., Jr. 2004.** Interaction of fire, climate and vegetation change at a large landscape scale in the Big Woods of Minnesota, USA. Holocene. 14: 661–676.

**U.S. Department of Agriculture, Forest Service [USDA FS]. 2005.** A strategic assessment of forest biomass and fuel reduction treatments in Western States. Gen. Tech. Rep. RMRS-GTR-149. Fort Collins, CO: Rocky Mountain Research Station. 17 p.

**U.S. Department of Agriculture, Forest Service; U.S. Department of the Interior [USDA USDI]. 2001.** National Fire Plan. A report to the President in response to the wildfires of 2000, September 8, 2000: managing the impact of wildfires on communities and the environment. Washington, DC. [Pages unknown].

**U.S. Department of Agriculture, Natural Resources Conservation Service [USDA NRCS]. 2008.** The PLANTS database. http://plants.usda.gov. (11 June 2008).

**Vale, T.R. 2000.** Pre-Columbian North America: pristine or humanized—or both? Ecological Restoration. 18: 2–3.

**Vale, T.R., ed. 2002.** Fire, native peoples, and the natural landscape. Covelo, CA: Island Press. 340 p.

**Vanderlinden, L.A. 1996.** Applying stand replacement prescribed fire in Alaska. In: Hardy, C.C.; Arno, S.F., eds. The use of fire in forest restoration. Gen. Tech. Rep. INT-GTR-341. Ogden, UT: U.S. Department of Agriculture, Forest Service, Intermountain Research Station: 78–80.

**van der Werf, G.R.; Randerson, J.T.; Collatz, G.J.; Giglio, L.; Kasibhatla, P.S.; Arellano, A.F., Jr.; Olsen, S.C.; Kasischke, E.S. 2004.** Continental-scale partitioning of fire emissions during the 1997 to 2001 El Niño/La Niña period. Science. 303: 73–76.

**Van Horne, M.L.; Fúle, P.Z. 2006.** Comparing methods of reconstructing fire history using fire scars in a Southwestern United States ponderosa pine forest. Canadian Journal of Forest Research. 36: 855–867.

**Vankat, J.L. 1977.** Fire and man in Sequoia National Park. Annals of the Association of American Geographers. 67: 17–27.

**van Lear, D.H. 1991.** Fire and oak regeneration in the southern Appalachians. In: Nodvin, S.C.; Waldrop, T.A., eds. Fire and the environment: ecological and cultural perspectives. Gen. Tech. Rep. SE-69. Ashville, NC: U.S. Department of Agriculture, Forest Service, Southeastern Forest Experiment Station: 15–21.

**van Lear, D.H.; Waldrop, T.A. 1989.** History, use and effects of fire in the Appalachians. Gen. Tech. Rep. SE-54. Ashville, NC: U.S. Department of Agriculture, Forest Service, Southeastern Forest Experiment Station. 20 p.

**Van Wagner, C.E. 1977.** Conditions for the start and spread of crown fire. Canadian Journal of Forest Research. 7: 23–34.

**Van Wagner, C.E. 1987.** Development and structure of the Canadian forest fire weather index system. Forestry Tech. Rep. 35. Ottawa: Canadian Forest Service. 36 p.

**van Wagtendonk, J.W. 1996.** Use of a deterministic fire growth model to test fuel treatments. Sierra Nevada Ecosystem Project: Final report to Congress, Vol. II. Assessments and scientific basis for management options. Water Resources Center Report No. 37. Davis, CA: Centers for Water and Wildland Resources, University of California: 1155–1166.

**Veblen, T.T. 2003.** Key issues in fire regime research for fuels management and ecological restoration. In: Omi, P.N.; Joyce, L.A., eds. Fire, fuel treatments, and ecological restoration: conference proceedings. Proceedings RMRS-P-29. Fort Collins, CO: U.S. Department of Agriculture, Forest Service, Rocky Mountain Research Station: 319–333.

**Veblen, T.T.; Kitzberger, T.; Donnegan, J. 2000.** Climatic and human influences on fire regimes in ponderosa pine forests in the Colorado Front Range. Ecological Applications. 10: 1178–1195.

**Vitousek, P.M.; Aber, J.D.; Goodale, C.L.; Aplet, G.H. 2000.** Global change and wilderness science. In: Cole, D.N.; McCool, S.F.; Freimund, W.A.; O'Loughlin, J., eds. 2000. Wilderness science in a time of change conference-Volume 1: Changing perspectives and future directions. Proceedings RMRS-P-15 VOL-1. Ogden, UT: U.S. Department of Agriculture, Forest Service, Rocky Mountain Research Station: 5–9.

**Vose, J.M. 2000.** Perspectives on using prescribed fire to achieve desired ecosystem conditions. In: Moser, W.K.; Moser, C.F., eds. Fire and forest ecology: innovative silviculture and vegetation management. Proceeding: 21$^{st}$ Tall Timbers fire ecology conference. Tallahassee, FL: Tall Timbers Research Station: 12–17.

**Wagle, R.F.; Eakle, T.W. 1979.** A controlled burn reduces the impact of a subsequent wildfire in a ponderosa pine vegetation type. Forest Science. 25: 123–129.

**Wahlenberg, W.G. 1946.** Longleaf pine: its use, ecology, regeneration, protection, growth, and management. Washington, DC: Charles Lathrop Pack Forestry Foundation. In cooperation with: U.S. Department of Agriculture, Forest Service. 429 p.

**Waldrop, T.A.; Brose, P.H. 1999.** A comparison of fire intensity levels for stand replacement of table mountain pine (*Pinus pungens* Lamb.). Forest Ecology and Management. 113: 155-166.

**Waldrop, T.A.; Brose, P.H.; Welch, N.T.; Mohr, H.H.; Gray, E.A.; Tainter, F.H.; Ellis, L.E. 2003.** Are crown fires necessary for Table Mountain pine? In: Galley, K.E.M.; Klinger, R.C.; Sugihara, N.G., eds. Proceedings of fire conference 2000: the first national congress on fire ecology, prevention, and management. Misc. Publ. 13. Tallahassee, FL: Tall Timbers Research Station: 157–163.

**Walker, J.; Peet, R.K. 1983.** Composition and species diversity of pine-wire grass savannas of the Green Swamp, North Carolina. Vegetatio. 55: 163–179.

**Wallin, K.; Kolb, K.T.; Skov, K.; Wagner, M. 2004.** Seven-year results of thinning and burning restoration treatments on old ponderosa pines at the Gus Pearson Natural Area. Restoration Ecology. 12: 239–247.

**Walstad, J.D.; Radosevich, S.R.; Sandberg, D.V. 1990.** Introduction to natural and prescribed fire in Pacific Northwest forests. In: Walstad, J.D.; Radosevich, S.R.; Sandberg, D.V., eds. Natural and prescribed fire in Pacific Northwest forests. Corvallis, OR: Oregon State University Press: 3–5.

**Waltz, A.E.M.; Fulé, P.Z.; Covington, W.W.; Moore, M.M. 2003.** Diversity in ponderosa pine forest structure following ecological restoration treatments. Forest Science. 49: 885–900.

**Ward, P.C.; Tithecott, A.G.; Wotton, B.M. 2001.** Reply–A re-examination of the effects of fire suppression in the boreal forest. Canadian Journal of Forest Research. 31: 1467–1480.

**Weatherspoon, C.P.; Skinner, C.N. 1995.** An assessment of factors associated with damage to tree crowns from the 1987 wildfires in northern California. Forest Science. 41: 430–451.

**Weaver, H. 1968.** Fire and its relationship to ponderosa pine. In: Proceedings: 7th Tall Timbers fire ecology conference. Tallahassee, FL: Tall Timbers Research Station: 7: 127–149.

**Weir, J.M.H.; Johnson, E.A. 1998.** Effects of escaped settlement fires and logging on forest composition in the mixed wood boreal forest. Canadian Journal of Forest Research. 28: 459–467.

**Weir, J.M.H.; Johnson, E.A.; Miyanishi, K. 2000.** Fire frequency and the spatial age mosaic of the mixed-wood boreal forest in western Canada. Ecological Applications. 10: 1162–1177.

**Weisberg, P.J. 2004.** Importance of non-stand-replacing fire for development of forest structure in the Pacific Northwest, USA. Forest Science. 50: 245–258.

**Weisberg, P.J.; Swanson, F.J. 2003.** Regional synchroneity in fire regimes of western Oregon and Washington, USA. Forest Ecology and Management. 172: 17–28.

**Weise, D.R.; Regelbrugge, J.C.; Paysen, T.E.; Conard, S.G. 2002.** Fire occurrence on southern Californian national forests—has it changed recently? In: Sugihara, N.G.; Morales, M.E.; Morales, T.J., eds. Proceedings of the symposium: fire in California ecosystems: integrating ecology, prevention and management. Misc. Publ. 1. Berkeley, CA: Association for Fire Ecology: 389–391.

**Weise, D.R.; Zhou, X.; Sun, L.; Mahalingam, S. 2003.** Fire spread in chaparral—"go or no go?" In: Proceedings, 5th symposium on fire and forest meteorology, and 2nd international wildland fire ecology and fire management congress. Orlando, FL: American Meteorological Society. http://ams.confex.com/ams/pdfpapers/65238.pdf. (23 January 2008).

**Welch, B.L.; Criddle, C. 2003.** Countering misinformation concerning big sagebrush. Res. Pap. RMRS-RP-40. Fort Collins, CO: U.S. Department of Agriculture, Forest Service, Rocky Mountain Research Station. 28 p.

**Wells, B.W. 1942.** Ecological problems of the Southeastern United States Coastal Plain. Botanical Review. 8: 533–561.

**Wells, P.V. 1962.** Vegetation in relation to geological substratum and fire in the San Luis Obispo quadrangle, California. Ecological Monographs. 32: 79–103.

**Westerling, A.L.; Gershunov, A.; Cayan, R.; Barnett, T.P. 2002.** Long lead statistical forecasts of area burned in Western U.S. wildfires by ecosystem province. International Journal of Wildland Fire. 11: 257–266.

**Westerling, A.L.; Hidalgo, H.G.; Cayan, D.R.; Swetnam, T.W. 2006.** Warming and earlier spring increase Western U.S. forest wildfire activity. Science. 313: 940–943.

**Weyenberg, S.A.; Frelich, L.E.; Reich, P.B. 2004.** Logging versus fire: How does disturbance type influence the abundance of *Pinus strobus* regeneration? Silva Fennica. 38: 179–194.

**Whisenant, S.G. 1990.** Changing fire frequencies on Idaho's Snake River plains; ecological and management implications. In: McArthur, D.D.; Rommey, E.M.; Smith, S.O.; Tueller, P.T., eds. Proceedings—symposium on cheatgrass invasion, shrub die-off, and other aspects of shrub biology and management. Gen. Tech. Rep. INT-276. Ogden, UT: U.S. Department of Agriculture, Forest Service, Intermountain Research Station: 4–10.

**Whitlock, C.; Skinner, C.N.; Bartlein, P.J.; Minckley, T.; Mohr, J.A 2004.** Comparison of charcoal and tree-ring records of recent fires in the eastern Klamath Mountains, California, USA. Canadian Journal of Forest Research. 34: 2110–2121.

**Wienk, C.L.; Sieg, C.H.; McPherson, G.R. 2004.** Evaluating the role of cutting treatments, fire and soil seed banks in an experimental framework in ponderosa pine forests of the Black Hills, South Dakota. Forest Ecology and Management. 192: 375–393.

**Willis, K.J.; Birks, J.B. 2006.** What is natural? The need for a long-term perspective in biodiversity conservation. Science. 313: 1261–1265.

**Wright, R.J.; Hart, S.C. 1997.** Nitrogen and phosphorus status in a ponderosa pine forest after 20 years of interval burning. Ecoscience. 4: 526–533.

**Yoder, J.; Engle, D.; Fuhlendort, S. 2004.** Liability, incentives, and prescribed fire for ecosystem management. Frontiers in Ecology and the Environment. 2: 361–366.

**Zedler, P.H. 1995.** Are some plants born to burn? Trends in Ecology and Evolution. 10: 393–395.

**Zedler, P.H.; Seiger, L.A. 2000.** Age mosaics and fire size in chaparral: a simulation study. In: Keeley, E.; Baer-Keeley, M.; Fotheringham, C.J., eds. 2$^{nd}$ interface between ecology and land development in California. U.S. Geological Survey open file report 00-62. Sacramento, CA: U.S. Geological Survey: 9–18.

**Zimmerman, G.T.; Neuenschwander, L.F. 1983.** Fuel-load reductions resulting from prescribed burning in grazed and ungrazed Douglas-fir stands. Journal of Range Management. 36: 346–350.

**Zimmerman, G.T.; Omi, P.H. 1998.** Fire restoration options in lodgepole pine ecosystems. In: Pruden, T.L.; Brennan, L.A., eds. Fire in ecosystem management: shifting the paradigm from suppression to prescription. Proceedings: 20$^{th}$ Tall Timbers fire ecology conference. Tallahassee, FL: Tall Timbers Research Station: 285–297.

**Zimov, S.A.; Davidov, S.P.; Zimova, G.M.; Davidova, A.I.; Chapin, F.S., III; Chapin, M.C.; Reynolds, J.F. 1999.** Contribution of disturbance to increasing seasonal amplitude of atmospheric $CO_2$. Science. 284: 1973–1976.

CPSIA information can be obtained
at www.ICGtesting.com
Printed in the USA
LVHW060315250820
664167LV00015B/1056